AMERICANS AND THEIR LAND

AMERICANS AND THEIR LAND

THE HOUSE BUILT ON ABUNDANCE

ANNE MACKIN

THE UNIVERSITY OF MICHIGAN PRESS ANN ARBOR

To Alex, Isara, and Isaiah,
who opened before me
the landscape of the heart.

Published in the United States of America by
The University of Michigan Press
Manufactured in the United States of America
∞ Printed on acid-free paper

2009 2008 2007 2006 4 3 2 1

A CIP catalog record for this book is available from the British Library.

Library of Congress Cataloging-in-Publication Data

Mackin, Anne.
Americans and their land / Anne Mackin.
p. cm.
Includes bibliographical references and index.
ISBN-13: 978-0-472-11556-3 (cloth : alk. paper)
ISBN-10: 0-472-11556-1 (pbk. : alk. paper) 1. Land use—United States—History. 2. Land tenure—United States—History. 3. Democracy—United States—History. I. Title.
HD191.M33 2006
333.30973—dc22 2006008014

First are the capes, then are the shorelands, now
The blue Appalachians faint at the day rise;
The willows shudder with light on the long Ohio:
The lakes scatter the low sun: the prairies
Slide out of dark: in the eddy of clean air
The smoke goes up from the high plains of Wyoming:
The steep Sierras arise: the struck foam
Flames at the wind's heel on the far Pacific.

—Archibald Macleish, "American Letter"

CONTENTS

PREFACE

My first impressions of the relationship between Americans and their land came during trips to my grandparents. Every few months, my mother, brother, and I piled into the car to drive two hours south from Arlington, Virginia, to Richmond.

There, along with the corn and tomatoes they grew in their ample garden, my grandfather and grandmother gave us their stories of growing up on farms at the turn of the last century, stories of their own grandparents and a parade of middle-class ancestors—farmers and clerics—that stretched all the way back to the early European settlement of Virginia. In my mind, the mists of family history and the summer morning mist on the Virginia landscape waft together, anointing the land with the aura of ancestry.

The first critique of the American landscape I can remember hearing came from my brother when we were teenagers. He had hitchhiked to Florida to see an Apollo launch and had hiked part of the way back on the Appalachian Trail, rambling through the Smoky Mountains before returning home. "What God has done in the South is beautiful," reported my brother, usually an avowed atheist, "but what man has done. . . ." He shook his head.

My brother's casual assessment describes many parts of the country and is at the core of contemporary criticism of the American landscape. The common answer is that the landscape of democracy, in which everyone gets a little plot, should not necessarily be beautiful but comfortable and affordable to the majority, like a polyester leisure suit. The answer to that answer is that the American landscape expresses not the triumph of democracy so much as the triumph of capitalism.

It only takes a little travel, which I undertook occasionally in the following years, to learn that, whatever the social, legal, and economic reasons for the appearance of our landscape, that landscape expresses uniquely American attitudes toward land, resources, and community. Sorting these attitudes, I've found, could be a life's work—or the subject of a book.

After a few years of graduate school spent studying landscape architecture and planning, I worked as a planner for the Commonwealth of Massachusetts. I spent two years driving the highways and back roads of Massachusetts en route to some of the communities my state office was charged with visiting. As on those early trips to Richmond, when our family car passed the cheap-looking motels and Civil War markers of Route 1, I had time to wonder why people built on the land in the way that they did. In fact, it was my job to wonder. Thus began my real education.

Chats with local officials, questions and comments at public meetings—all these revealed the different economies and personalities of the communities I visited. I found towns whose fortunes had risen and fallen with ancient industries: whaling, farming, boot or textile manufacturing. Some towns were forgotten and pristine; some forgotten and derelict. Some had been overrun by modern development and were now barely recognizable as distinct places amid the sprawling development around them. Some little towns threw tremendous energy and goodwill into the struggle for self-improvement, and some thumbed their noses at the idea of accepting so much as a street tree from state government. Others still—typically the affluent communities with much to protect—operated sophisticated planning agencies of their own with powerful boards overseeing zoning and other aspects of development.

In all of these communities, the compelling force that shaped the land did not arise in any of the programs of state government—designed to preserve some of the natural beauty and cultural history of the Massachusetts landscape—but with the powerful forces of commerce. Compared to the robust pressures of real estate development, the efforts of government appeared relatively fragile and ineffectual.

The cultural forces shaping the landscape became even clearer to me after I left my state job, returned to writing part-time (mostly about planning), and had children. For the first time, the commercial strip became part of my life as I strapped my first little toddler into her car seat and headed for Toys R Us and Chuck E. Cheese on a rainy day.

Suddenly, I owned my own car—instead of sharing or avoiding them—and by the time the children were preschoolers, I was using the drive-through lane at McDonald's when caught off-guard by squalls of hunger from the backseat. I had to admit the developers of sprawl were on to something. With the marketer's honed instincts about the lifestyles of middle-class families, the many corporations whose logos lined the strip had anticipated my needs and the needs of other parents and had answered with the lowest common denominator.

Although my forays to the strip lasted only until my second child entered school, I understood my collusion with the developers of junk environments. The careless, disposable quality of this hawking landscape interested me, as did the easy plasticity with which it was molded, and the abundance of land that permitted it. Surely, the cultural attitudes that created this landscape resulted from centuries of enjoying land so abundant, and government strictures so loose, that the land was used as casually as any modern, throwaway plastic utensil might be.

With the vague idea of a book in mind, I began years of on-and-off research on the history of settlement patterns and the distribution of land in America. Through all of this, I have come to believe that we cannot fully understand our modern society and its relationship to the land and resources on which it relies without understanding Americans' historic relationship to their land—a resource once so abundant as to seem limitless. But, as many have come to realize, what abundance set in motion, it cannot sustain. The cultural changes that we see in our modern society are brought about in part by a dramatically different ratio of people to resources than that experienced by our distant ancestors. It is important, I think, to make these changes as consciously as possible and to understand their causes and consequences, as well as the consequences of failing to make certain choices that we have not yet been able to make.

INTRODUCTION

What is the nature of the relationship between a people and the land they occupy? Cultural anthropologists tell us that different cultures arise from different environmental circumstances. The land makes the man and the woman. The great religions of the world, differing political and economic systems—all these are responses to the provisions of nature. Culture is, in a way, a by-product of the historic human effort to claim and divide resources. Culture is the complexity brought by humanity's other gifts—the spiritual, cerebral, psychological—to the simple struggle for survival.

If we borrow the cultural anthropologist's lens to examine the historic relationship of European Americans with their land, we find many complexities. Ours is a transplanted culture—an Old World culture planted in a new, uncrowded world of seemingly limitless fertile land. But the beaching of the first European shallop on American shores initiated an alchemy of nature and culture that produced a uniquely American society. With the motherland a two- or three-month sea voyage away, and with enough land and other resources to spread out and start anew almost continually, attitudes toward all traditions—including social conformity, civic responsibilities, individual freedoms, class and wealth, and the role of community—began an organic transformation.

But this is not the end of the story. People work the land while the land works on them. After four hundred years of working American soil and transforming American resources into a vibrant economy, we—the descendants of those first European Americans and the many immigrants who have arrived since—find ourselves in a different land, a more crowded land of diminished resources; in short, a different environment. And, not surprisingly, these new environmental conditions have begun to influence our culture and its institutions, bringing about further cultural changes, although these are little examined or acknowledged.

Cultural transformation is a slow business. As more people or potential uses compete for the same or fewer resources, our nation's social and political institutions sort winners and losers differently. The government no longer gives away free land, for example, but it does subsidize many aspects of our daily life and many of our industries, such as transportation, home building, and some agricultural production, to name only a few.

As competition for resources increases, conflicts increase, even amid our tremendous affluence. These conflicts fill our headlines today. The cry of "Not in My Backyard!" greets many development proposals. But other issues nag us from the headlines: Will we use our forests for economic purposes or for human enjoyment and wildlife? Who gets western water—farmers or city dwellers? Who gets to profit from the distribution of that water? Is it acceptable that our development patterns create sprawling settlements that are difficult to administer, consume ever more fuel and time to navigate, and are beginning to consume economically productive land such as farmland, timberlands, and land atop oil fields? Should a rancher own the oil, minerals, and water under his or her land, or is it legitimate for the rest of society to take what it needs from belowground? Should that rancher have access to public land for grazing cattle? How or why should we provide affordable housing in most of our major metropolitan areas where housing costs are spiraling out of sight?

These issues require us to measure the individual rights and benefits associated with land against the interests of all of us together—the community. And it is that central calculus that gradually changes as a neighborhood or a nation becomes more crowded. The way one person uses his or her land comes to affect more people more dramatically. The history of that contest between individual and community needs in the division of resources—traditionally mediated by policy and law—is, for that reason, a theme of this book.

That contest shaped American government. Without wealth but with a vast store of land, and a population clamoring for it, the federal government in its first century dealt extensively with land distribution. The government subsidized early transportation investments to support the commercial vitality of different regions. By the beginning of the twentieth century, the federal government began to oversee water distribution to support western agriculture and western growth. The federal government began to promote zoning in the 1920s to help reduce land use litigation in growing metropolitan areas and suburbanizing rural areas. And, since the 1960s, Congress has passed many acts of legislation aimed at protecting the public health in the face of deteriorating air, soil, and water quality, as well as many acts of land conservation. Government also directs resources in subtler ways, through supports for specific industries and for commerce in general. So the question now facing us is not whether government should oversee the apportionment of resources—it already does—but how.

Because the American legal and political systems, along with our attitudes toward property, owe so much to the English colonies and the precedents on which they relied, the book begins there. Certain basic injustices associated with our early use of land—the way we acquired it from the Native Americans and made it productive through the use of slave labor—are such large topics, and so well covered in other books, that I have touched on them only tangentially here, as they relate to our attitudes toward land and the development of our economic system. Chapters 8 and 9, however, discuss slavery's aftermath, in which a large population of socially and economically displaced African Americans vied for a place in northern industrial cities.

The relationship between people and their land is indeed complex, and the discussion in this book is only a beginning, intended to bring further attention to these issues. Land is only one of the resources we share, but it became the focus of this book for several reasons. Land is home to most other resources, including fresh water, minerals, oil, gas, and plant life. Since it cannot be imported or acquired except through territorial expansion, the stresses of sharing land have become more obviously pressing than the stresses of sharing other resources—except oil, gas, and, increasingly, water.

There is, of course, no agreement on the actual extent of most of our resources. Even a simple calculation such as the area of the United States does not tell us how much land is suitable for home building, with adequate water supplies, or how much agricultural land or other economically productive land can be safely sacrificed to home building as our population grows.

For that reason, an inventory of resources is not necessarily the most revealing indicator of America's future. We are the most revealing indicator of that future. Long before we consume our last drop of water or oil or our last acre of land, we will have become a different society. Will that society pursue more or less equitable ways to apportion remaining resources? How will we confront the pressures that scarcity places on a democracy? How will we decide who gets to use the water and who gets to foul the air? The decisions we have already made are a good guide to our future, and it is the history of those decisions that this book attempts to trace.

SECTION 1

Perpetual Growth in the Land of Abundance

Many men will have living on these lands:
There is wealth in the earth for them all and the
wood standing

And wild birds on the water where they sleep:
There is stone in the hills for the towns of a
great People . . .

—Archibald MacLeish, "Empire Builders"

CHAPTER 1

PERPETUAL GROWTH IN THE LAND OF ABUNDANCE

It is odd to watch with what feverish ardor the Americans pursue prosperity and how they are ever tormented by the shadowy suspicion that they may not have chosen the shortest route to get it.

—Alexis de Tocqueville, *Democracy in America*

Nothing has shaped the American character as fundamentally as the sheer abundance of land and resources that met the European colonist, that met the westward-moving settler, and that still appears to meet the American suburbanite pressing into the exurbs. Abundant land supported the frenzied colonial race to procure and use resources—the race that became the large and vigorous American economy. Abundance has supported the prosperity that underpins America's democratic attitudes, all founded on the belief that there can be enough for all. Plentiful land fueled the individual fantasies of fulfillment that drew settlers across mountains and prairies and that have translated into a political preference for strong individual rights. Abundance has also supported collective utopian fantasies from the Puritans' "city upon a hill" and other religious communities to model industrial towns and all the way to the New Deal and the Great Society of the twentieth century.

Abundant land supported social and economic mobility as new opportunities redistributed wealth. Not only was land abundant enough to increase landownership rates, but a shortage of labor to support the continual refinement and growth of colonial settlements led to higher wages for laborers and tradesmen. Higher education became more accessible in New England than in the Old World, with early Harvard students paying their board and tuition in livestock or sacks of grain.[1]

Plentiful land allowed people to live farther apart within their own communities—which Americans chose repeatedly to do—or to break away altogether to follow the frontier as it migrated across the country, permitting the breakdown of restrictive social traditions as well as some good ones, such as the sense of civic obligation that we are currently trying to resuscitate. The space, privacy, and independence afforded by abundant land have become American luxuries, viewed by many as necessities. The virtues we prize are the virtues of the settlers who first grasped those luxuries: independence and industry.

Plentiful land underlay the national prosperity of the United States as well as the individual prosperity of her citizens. Land paid soldiers.[2] The sale of federal land granted to states built roads, canals, railroads, and colleges to further the agricultural and applied sciences. The abundance of land and the need to develop policies and practices for distributing it shaped the federal government around democratic principles of land distribution. Abundant land allowed the young federal government to court the scattering settlers with liberal land distribution policies: at first cheap land and finally, in 1862, free land through the Homestead Act.

As the settlers swept over Native American homelands, the early federal administrations budgeted heavily for the settlers' protection: an army to decimate and resettle the western Indians and to protect America's borders from the British, Spanish and, finally, Mexicans. The American right to enterprise was well secured by settlers even before the American Industrial Revolution gave enterprise its more modern connotations.

The vastness of the continent prompted the government to undertake or partially underwrite transportation initiatives such as canals, highways, and railroads in order to promote the economic welfare of the nation generally and of various regions particularly by connecting them to larger markets. This, in turn, helped foster the national zeal for technological advancement. The growth of larger mining and railroad projects—in particular the transcontinental rail lines designated and subsidized by the federal government—required the creation of a new fiscal entity: the heavily capitalized corporation. These corporations had to raise enormous capital far in advance of obtaining any profits. The most successful of these, in turn, wielded influence that changed the American political and economic systems—changes manifest in our lifestyles and landscape, which we are now trying to understand and meet.

The rapid growth that resulted when the land-poor, technologically sophisticated Anglo-Europeans of the seventeenth and eighteenth centuries arrived in the bounty of a more thinly populated "New World" has continued. The juggernaut of American growth has changed our landscape and taxed resources. Growth has given us a new and starkly different land in which to make our twenty-first-century home. And public attitudes are slowly responding to the new conditions of a more crowded land of partially depleted resources. And yet, the urge to grow, to produce, to gain, remains a powerful American cultural trait—arguably vital to our economy.

Economic growth is a common enough trait of democratic societies, and population growth is usually one of its engines, but American history offers the most striking example of these traits in tandem. The impetus of population growth and the goal of economic growth spurred the countries of Europe to sponsor New World colonization in order to obtain raw materials and wealth. Rapid growth, with occasional setbacks, characterized the American colonial period and the long frontier era and saw the country settled at a breathtaking pace that can be contrasted with the concurrent but slower settlement of Canada, still under British dominion.[3] American economic growth is still, or has been until recently, the envy of the world.

Most Americans consider growth—in the individual form of their own economic gain—to be vitally important: growth of the stock market, growth of job opportunities, growth of salaries, and the expansion of conveniences and luxuries that make up the American lifestyle. And where has this growth led us? To a high standard of living in a changed landscape of diminished resources. In this *new* New World, many of our cultural traits, nurtured by the abundance of a younger land, are not entirely at home.

At the moment, our ever-accelerating consumption of resources shapes public experience, public attitudes, and government policy in ways that we can all see. There are changes in the posture of citizen toward government and government toward citizen, expressed mainly through regulation and regulatory battles. There are changes in the attitude of citizen toward fellow citizen. In addition, as levels of some American resources fail to keep up with the consumerist habits of our growing population, our dependence on the resources of other nations—oil being the most popular example at the moment—has wrought profound changes in international relations and American foreign policy—especially trade and military policies—that reach into the daily lives of all Americans.

When we want to evaluate ourselves, it is usually best to look through other eyes—an act of imagination. America has had many astute observers. The most relevant to our purpose may be those who took early note of America and pondered from a distance the future we now inhabit. All European and all born within forty years of each other during the brewing of the American Revolution and the period of America's young nationhood, they were drawn, to varying degrees, to contemplate the American experiment. Malthus noted that "the happiness of the Americans depended much less upon their peculiar degree of civilization than upon the peculiarity of their situation, as new colonies, upon their having a great plenty of fertile uncultivated land."[4] In this observation, he echoed remarks made years earlier by Jefferson's brilliant, Swiss-born secretary of the treasury, Albert Gallatin. Later still, the historian and social critic, Baron Thomas Babington Macaulay visited America. Brilliant and earnest Macaulay helped defeat slavery in the West Indies, but he distrusted democracy. He charged that American political institutions would not be truly tested until their supply of land ran low. And, of course, Alexis de Tocqueville endeared himself to generations of Americans by dedicating to us over seven hundred pages of observation and analysis. Among other qualities, de Tocqueville characterized our restless bent on prosperity, our love of industry over the arts, and our hungry patronage of the mass-produced trappings of wealth—the patterns of consumption that would reduce the abundance of which Malthus, Gallatin, and Macaulay wrote. In fact, if we could invite this august group for a follow-up visit as a committee, they would find themselves in a land that boasts the most extravagant rate of consumption ever conceived on earth by nonroyalty.

I like to think of this eloquent and incisive committee in modern California, standing on the Hollywood Hills. None has visited California, and de Tocqueville, perhaps, would be pleased to view a part of America that he missed on his first tour—when it belonged to Mexico—a part that represents the most recent evolution of American culture. Like Balboa's party—centuries earlier and much farther south—de Tocqueville and his companions are perched on the western coast of the New World, viewing an unimaginable vastness—and on the verge of understanding their discovery. But they are not seeing the Pacific. They are looking at modern Los Angeles—427 square miles within its city limits but sprawling 4,000 square miles over its greater metropolitan area—down to meet the outskirts of San Diego's suburbs. Perhaps, at this moment, de Tocqueville turns to us like Keats's Balboa-Cortez, eyes wide "with a wild surmise."[5] Now we must explain to our guests that, while Los Angeles has come to symbolize modern American sprawl, its metropolitan core happens to be smaller than those of Jacksonville, Oklahoma City, and Houston—and barely larger than those of six other major American cities.

Malthus will want to know that at three hundred million our national population is large but wealthy by eighteenth-century standards and that the birth rate is low, about 1.4 per 100, but augmented by immigration, confirming his later theory that slower population growth is tied to higher levels of wealth. He will have to look longer and more closely to test his other theories.

Our illustrious group is looking down on the result of economic growth that—on paper and on the ground—dazzles foreign visitors. That economic growth has been supported by rapid population growth that, having slowed, is sometimes dismissed as negligible. However, our tiny rate of increase currently adds nearly three million people to our landscape annually. Someone born in 1960 now shares his or her landscape with an additional seventy-six million people.[6]

De Tocqueville toured an America with just under thirteen million residents. There are now thirty-four million people living in California alone, the state where we have positioned our group. In 1831, the time of de Tocqueville's visit, 90 percent of Americans still lived in rural areas, supporting the still-agrarian economy. Cheap land of good quality was readily available in many regions. Now, over 80 percent of Americans live in urban areas—mainly suburbs—creating stiff competition for land in the metropolitan areas that hold most of us.

From the modern hills overlooking Los Angeles, our guests would be looking down on an endless cityscape cloaked in the pollution haze of industry and automobile. In 2001, the Los Angeles Basin exceeded federal health standards for air pollution in an eight-hour period on one hundred days.

California is famous for its congested freeways, but all across America we love our polluting automobiles. They allow us to live in the suburbs that so many of us seem to love, sprawling steadily beyond the reach of public transit. And land is still the stuff on which we build our dreams. Sixty-eight percent of us own our own home, even if some are small apartments, motor homes, or in poor repair.

As a group of individual consumers, we spend over seven trillion dollars annually on those dreams—everything from health care, home, and transportation to clothing, food, and tobacco. Every year, each of us consumes about four times the resources consumed by the average citizen of the world—over twenty times that of our neighbors in poor countries.

Supported by new technologies of energy generation, transportation, communications, and manufacturing, with endless retail outlets, each of us takes up about twice as much developed land as the average U.S. resident of 1930—our parents or grandparents. In 1930, America had less than half an acre of urban land per person in the United States. For every American in 1992, America had over an acre of urban land.[7] To use the words of a recent planning bulletin, "Most metropolitan areas are consuming land for urbanization much more rapidly than they are adding population."[8]

Our committee would be looking at the subdivisions that ate up American land at the rate of two million acres a year in the 1950s and 1960s, along with other kinds of development. They would be looking down on superstores—with footprints larger than Notre Dame Cathedral or Westminster Abbey—part of the development recipe that gobbles up over two million acres a year today.[9] Counting on the average superstore footprint of sixteen acres, including parking and buffers, we can calculate that Home Depot alone consumed approximately three thousand acres in the year 2000, when they opened a new store every two days.[10] McDonald's opens five new stores a day worldwide, with fifteen thousand already in North America. McDonald's, according to Eric Schlosser's *Fast Food Nation*, "is the largest owner of retail property in the world," earning more from collecting rent than selling food.

Although de Tocqueville and the others may not be able to see it from their present lofty perch, as our stores have expanded and multiplied, and the products on their shelves have multiplied, the debt of the average American has also multiplied, and even the bulk of the average American has expanded to accommodate the myriad variety of new commercial offerings that include fast food and glossily bagged or boxed snacks—and so have our homes and settlements.

The dreams of Americans—homes, schools, industries, services, stores, recreational facilities, houses of worship—are connected by cars—rivers and rivers of them. The once legendary traffic jams of Los Angeles—now common to most major American cities—also represent economic and population growth: the growth of the automobile industry and its lobby and the increasing number of drivers, the increasing number of miles driven per individual, and the demand for more cars per family. If our committee happens to be watching during any of the six hours of the modern "rush hour" (three hours in the morning and three in the evening), they will not find many of the cars moving. According to the same report that recently calculated our longer rush hours—the 2005 Texas Transportation Institute's *Urban Mobility Report*—each resident of Los Angeles annually spends about ninety-three hours—or eleven and a half workdays—stuck in traffic, and most of us were stuck in traffic three times longer in 2003 than in 1982. In less densely populated areas, where more development now occurs, the average American spent five times as long stuck in 1999 traffic as in 1982 traffic.

Invisible to our committee would be the fuel used by these cars. Transportation uses account for the majority of oil consumed by the United States at an ever-accelerating rate. The words of two Princeton economics professors, William J. Baumol and Wallace E. Oates, help to dramatize accelerating global fuel consumption.

> It has been estimated, for example, that in the first two decades of the twentieth century mankind consumed more energy than it had used in total over all the previous centuries of its existence. During the following two decades, we again employed more

> power than in the totality of the past (including the part of the twentieth century that preceded it). Moreover, a similar statement has held for each twenty-year period since then. There has been a dramatic increase in recent energy consumption and the mind boggles at the future demands that an extrapolation of these exploding trends in demand suggests.[11]

In that frenzied consumption, America is the leading customer, using just over three times as much oil annually as the nearest competitor, China.[12] Much of the oil we use goes to the automobiles that convey us and our goods and materials around our bloated settlements. Cars and trucks not only use fuel extravagantly, but they reduce another resource: air. A large share of American air pollution originates with automobiles.

Malthus, however, would have tired of hearing about wealth by now. Accosting a passerby, he is taking directions to South Central L.A. and Watts, to see if he must revise his predictions of famine, disease, and war as checks to rising populations. Macaulay starts to follow, but his suspicion of commoners makes him hesitate. He inquires at a safe distance whether the "Jeffersonian polity" is anywhere near the "fatal calamity" he predicted for it. What is the unemployment rate here? How close to Europe's population density has America come? How much cheap or free land remains available in America?

De Tocqueville and Gallatin might still be drinking in the sights and sounds of this intriguing new country and beginning a discussion. They would not stay long on the heights either before wanting to see and hear the people of this strange new world. De Tocqueville, in particular, is not a counter of cars but a reader of desires. He must see a newspaper, hear a public meeting. Were we to accompany our guests in their descent, we would want to guide him to these sources. He must see how far democracy has succumbed to the seductions of luxury. We would want to help him meet a coffee sipper at Starbucks, a shopper at Wal-Mart. We would want to coax from our guests their initial impressions and offer, perhaps, our own observations about the culture that is now reconfiguring in the land of perpetual desire, perpetual growth, and diminished resources.

But, just as they would want to, we must go and see for ourselves. We can only imagine whether Malthus would, during such a visit, recalculate his theory of the disasters of overpopulation as we have done over the centuries of technological advance and imported resources. Or would he see subtle signs in American culture of the limits ahead: not pestilence so much as limited access to health care for the 46 percent of Americans without insurance? Not starvation but poorer nutrition, poorer health, and a shorter life span for the poor. And slow-burning wars for oil fought mainly by the poor. Would de Tocqueville take pride in the accuracy of his earlier insights, and what would he add? Would the Swiss-born Gallatin still admire his adopted country?

We cannot know. We will have to guess and make our own judgments. So, keeping in mind the images of urbanization and growth we have just witnessed with our committee of visionaries in mind, we embark in their stead to explore the cultural changes that have accompanied that physical transformation. And, before beginning, we should review a few striking clues.

Our attitudes toward growth and property rights are changing. Just as American cities and towns once clamored for growth—for the canals, railroads, or industries that would bring prosperity—many cities, towns, and counties around the nation now clamor to slow growth. Although even crowded locales still court businesses and other sources of tax revenues, their residents are often embattled, fighting congestion, intrusive land uses, and the rising taxes that pay for growth. Their weapons are increased land-use regulation and increased public oversight of development projects, signaling the public's desire to make community interests a greater factor in land-use decisions.

Increased regulation and resistance to growth, in turn, have aggravated developers and property rights proponents. The property rights movement opposes land use restraints with passion and, sometimes, with money from the real estate and construction industries. To be fair, residents whose sole wealth lies in their homes—such as elderly or long-time residents whose finances have been drained by property taxes driven up by rapid growth—sometimes camp with this movement. The property rights movement represents the individual's interests against those of the community, enjoying both recent victories and defeats. In the words of Professor Harvey Jacobs, in his recent book, *Private Property in the 21st Century*, "an era of social conflict over environmental values and resources is upon us, and one of its expressions will be heightened conflict over property rights."

The depletion and degradation of resources has begun to affect the lives of ordinary Americans. Is a reduced supply of resources beginning to erode our tremendous affluence? First, we face mounting hazards to public health. In addition to the public health hazards of fouled air, pesticides and other chemicals such as mercury that cause cancer and birth defects have materialized in our food supply. With a higher population and reduced land supply, home prices and rents consume a greater share of the American income, reducing the amount people can spend on other needs. Finally, as population has increased and the nation has urbanized, individuals have lost direct access to the resources that once provided their ancestors with food, fuel, and cottage industries.

The wealth gap between rich and poor is widening. Economic historians agree that frontier conditions—abundant resources available for the taking—promote economic equality.[13] But as resources are taken up, the wealthy and well connected have an advantage in obtaining what's left, for the obvious reason that, as a resource diminishes, its price rises and competition for it intensifies. If technological innovation restores a measure of access to that resource, those that control the technology, typically large corporations, are the ones that control the resource. This trend partly explains the growth, since the 1860s, of the large, heavily capitalized corporation and its greater role in our economic and political life.

As different interest groups struggle over resources, regulation increases to mediate the resulting conflicts. The consumption of land and resources; the resulting pollution, shortages, and conflicts over land uses (and the property rights adhering to them) have all obliged government to regulate more. The creation of the Department of Energy (and the many earlier agencies from which it evolved), the creation of the Environmental Protection Agency, the establishment of zoning earlier in the century, and the body of land use litigation throughout the century are all part of this trend. The economic changes that have accompanied the intensifying consumption of resources have also inspired legislation and regulation, from the antitrust laws of the late nineteenth century to modern laws governing the power industry and agricultural subsidies.

Although all resources have economic value, water is unique in being understood as essential to human life. The value of land without a dependable water supply—either for growing crops or slaking thirst—is nil. As water becomes scarcer and more precious, there are three immediate results. First, shortages occur and conflicts between different interest groups arise—notably between urban residents who want to drink water and farmers who want to water crops. Second, water prices rise. And third, large private corporations move into the increasingly lucrative field of water supply. Though local and state governments have been regulating water supplies for centuries in America, and the federal government has been doing so for over a century, recent shortages and their repercussions invite increased government intervention to manage the public water supply, and such intervention, in the form of regulation, has occurred and is intensifying.

Increased government regulation runs counter to the American pioneer spirit—the notion that America is a nation of government-averse individualists whose personal industry is spurred by government's willingness to let them do as they please. The fact that—despite political and regional differences on the subject—we, as a nation, have recognized the need for government protection from the hazards of pollution and resource shortfalls indicates our growing concern for our situation.

Most of us are no longer able to see the connection between our daily activities and the environmental change they cause. Unlike the hunter-gatherer or early farmer who knew where his food came from and where his waste went, we live in a world of clean, modern conveniences that consume resources and dump our waste invisibly. When we wake up in the morning and turn on a few lights, we do not see the coal burning in the next state that generates our electricity—or the vast infrastructure that brings us that electricity. In the bathroom, we do not see the reservoir from which our water supply comes, or the treatment plant to which our waste water travels, or the pollution in the rivers to which they return. We do not see the factories that produce the cosmetic products with which we wash, shave, and primp in the morning or those that produce the plastic, cardboard, and cellophane that package those products. We do not see the resources those factories use or the poisonous wastes they dispose of. Similarly, the clothes we reach for have been manufactured out of sight and cleaned by the invisible and toxic processes of the dry cleaner or by our own washing machines, consuming more electricity and gallons of fresh water, which are then conducted, with the dirt and detergent pollutants, to a nearby water body or treatment plant.

When we go into the kitchen to make breakfast, we do not see the many factories that have made our kitchen appliances, building materials, or the packaging in which our food comes—or their noxious efflux into water and air. We do not see the processing plants that handle our food or manufacture its many artificial ingredients—or their consumption and pollution. Again, the electricity or natural gas that fuels the refrigerator, the microwave oven, or the stove seems to arrive like magic, heralded only by a low hum that means burning fuel and billowing smoke many miles away. Somewhere in the northern woods, paper plants have used the marvels of toxic chemicals, electric power, and fossil fuels to change a forest the size of a small country into the morning paper that arrives on our doorsteps. When we drive to work, we are not really aware of the vast quantity of pollution released into the air all over America during the lengthening morning rush hours.

Welcome to the landscape of Dorian Gray. Like the famous portrait, our landscape shows our vices while our homes and countenances remain clean and bright. Sophisticated modern technologies hide our waste from us—along with our consumption and our culpability. So, while there is more of an outcry against the harmful effects of pollution, there is, simultaneously, a lack of understanding of how pollution is created, how it pervades our environment, and whether it exists at harmful levels, that prevents a national consensus.

The solution is part of the problem. As resources become scarcer, technological innovation often helps stave off shortages and price increases. Technological advances have heightened agricultural production, for example, and made it possible to extract less accessible deposits of oil and minerals.

And yet, the technological innovation and eagerness for which America is renown have had unforeseen consequences, ruining some resources while harvesting others. Modern mining, for example, employs new technologies to harvest inaccessible ores. These technologies lay waste to thousands of acres of land per site. Their machineries use fuel, and their refining processes leave toxic wastes. The extraction and transport of oil to fuel much of our technological advance results in pollution and oil spills. Even farming practices, as large-scale, mechanized agriculture has increased, have produced large-scale pollution.

Another unforeseen side of our escalating technological sophistication is the way it favors educated workers in our society, as shown in Frank Levy's *The New Dollars and Dreams.* The salary gap between a convenience store cashier and a white-collar worker in the technology industry is vast.

We might ask why this need for perpetual growth in the land of abundance. It isn't just population growth, obviously, since our standard of living keeps rising, our luxuries and choices of products continually increase. In a long forgotten article from 1974, economist Robert Heilbroner offered an explanation. He said that capitalism's "strong tendency to expand output" serves three purposes: "It expresses the drives and social values of its dominant class." It avoids the danger of a glut (which would cause a loss of profits and a cutback in production with further repercussions such as job losses). And lastly, it accommodates our constant striving "for larger rewards."[14]

These cultural attributes that fuel and complicate our economic growth are nearly part of the air we breathe. They have propelled the striking physical changes wrought by growth in America, the changes we have tried to consider through the eyes of Malthus, Gallatin, Macaulay, and de Tocqueville. And to fully understand them we must, like de Tocqueville, travel the nation and also travel its history.

We must go back to the moment when European culture began transplanting itself to the New World. And we must look through the eyes of an earlier witness.

Instead of the Pacific coast, we must find the Atlantic coast: beachhead of the Anglo-European culture from which the first American colonies grew. Northeast, then, to New England, and to the late sixteenth century. At the very eastern edge of Cape Cod, we will join a lone Nauset watching from a bluff as a small speck on the horizon grows gradually larger. And, knowing the onslaught that is about to begin, we might wish to offer him a few words of small comfort. We might wish to tell him that we have learned something now of tides of immigration, of continual growth pushing earlier residents onto smaller parcels of land of lower and lower quality, of environmental degradation and the disappearance of sacred places. We might wish to whisper, as the boat on the horizon draws nearer, that even *we* did not get to keep that lovely land.[15]

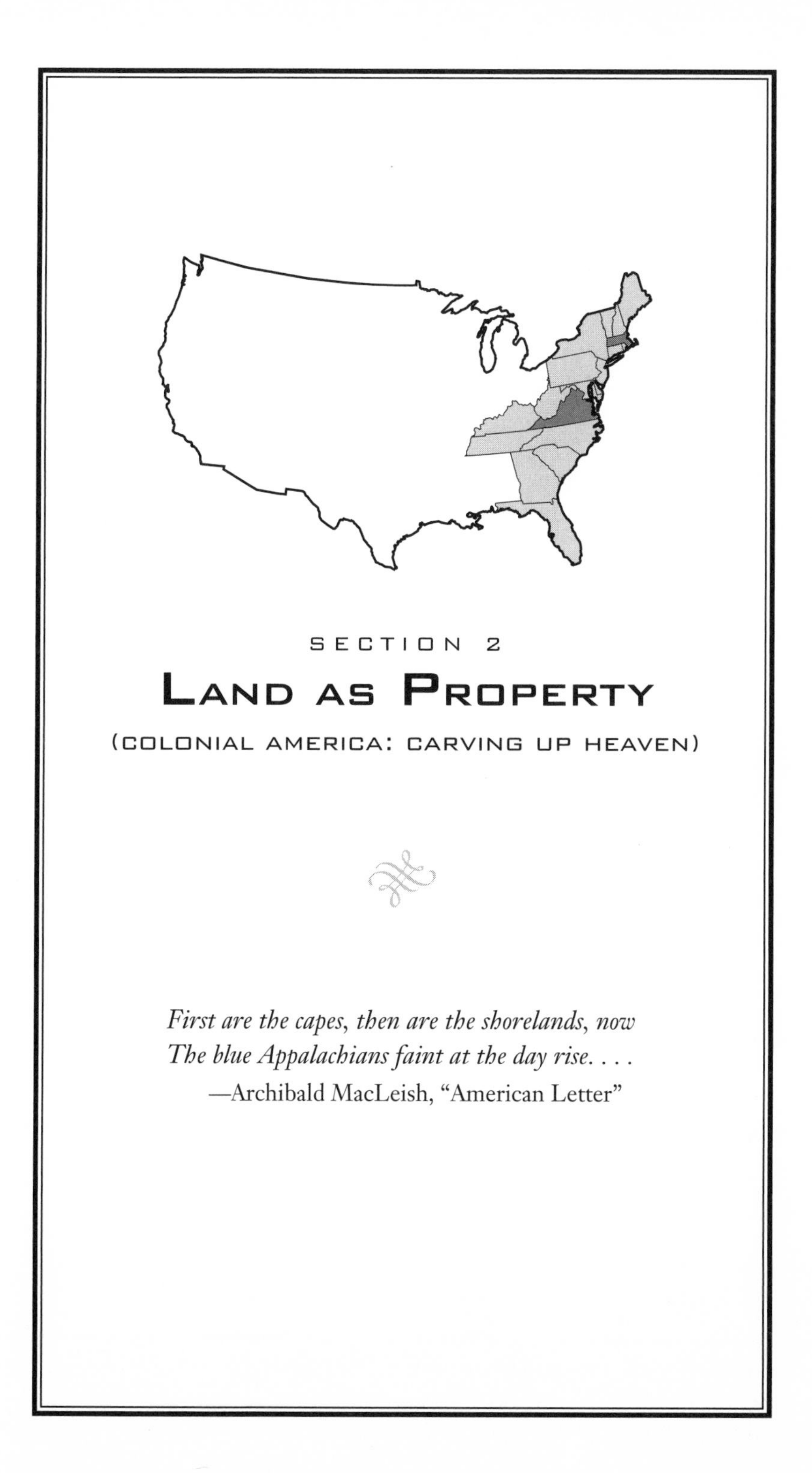

SECTION 2

Land as Property

(COLONIAL AMERICA: CARVING UP HEAVEN)

First are the capes, then are the shorelands, now
The blue Appalachians faint at the day rise. . . .
—Archibald MacLeish, "American Letter"

CHAPTER 2

LAND: THE EARLY BIRD SPECIAL

James, by the Grace of God, King of England, Scotland, France, and Ireland, Defender of the Faith . . . gives and grants to the Council established at Plymouth in the County of Devon, for the planting, ruling, ordering, and governing of New England in America, and to their successors and assignes for ever: All that parte of America lying and being in breadth from forty degrees of the said northerly latitude inclusively, and in length of and within all the breadth aforesaid throughout the maine landes from sea to sea.

—Royal Charter granting New England
to the Plymouth Company

At the eastern shores near the main Nauset village, the sands of Cape Cod stretch away into the Atlantic like history stretching into the dim past. As those sands shift in the currents beneath the waves—now plied by pleasure craft and fishing boat—they form and reform shoals and sandbars where the wrecks of hundreds of deep draught vessels lie, driven into the shallows in storms and pounded apart by the waves. The Nauset must have gathered an occasional strange cargo on the beaches after a storm, as Cape Codders did later in the seventeenth and eighteenth centuries when the wrecks became more frequent and the cargoes richer.

There is a modern bustle to town life on Cape Cod now, but on a clear, sunny day, the tourist can find a good east-facing bluff, shade his or her eyes, and peer out over those waves and into the past. For a moment, anyone can become a Nauset, watching, in the mind's eye, a European ship draw nearer over the sparkling waves.

The tourist needs a higher vantage point—say, atop Cadillac Mountain on Mount Desert Island, Maine—to push the horizon back beyond the pleasure boats to the curving silver-blue line where only a transatlantic voyage or a long hunt for commercial quantities of fish would take the adventurer. From this rocky perch, the ocean appears to grow close to its true, vast size. The waters at the horizon belong not to any continent but to the ocean alone. There, nations and centuries fall away and human life occupies a less significant place among the elements. One can imagine a tiny speck breaking the horizon—an ancient European ship nosing its way through the waves with a scurvy-plagued, ill-paid crew, nearly mad from eight to twelve weeks of seeing only sea and sky, indifferent to their fate. On this bright blankness, Europeans reached a New World—fishermen, whalers, traders, explorers, pirates, colonists, soldiers—Europeans and Americans of earlier centuries.

Whose ships lie under the waves of the New England and Canadian coasts? If we could have watched four or five hundred years ago, not from a bluff but from the shifting haunt of gulls, we would have seen the vast sparkling arena on which a bustling traffic of ships moved in search of new resources, their wakes the marks on a dazzling accounting ledger. Certainly, by the early sixteenth century, European fishermen could be found drying their catch on the craggy shores of Newfoundland. It now seems likely that fishermen and whalers beat Columbus to the New World and Cabot to North America. In 1535, Cartier found not only Basque ships when he toured the coast of Canada's Maritime Provinces but land forms already named by French fishermen, occasionally peopled by descendants of both French and Indians.[1] A British observer at the Grand Banks in 1578 counted 350 ships, predominantly Spanish and French with about 50 British.[2]

As the fish of the western Atlantic entered European fishing nets, they entered Europe's economy. Through expanded trade routes, that economy was rapidly absorbing raw materials and goods from various parts of the world. The speculative energy of Europe's economy (so familiar to us in our own economy) focused on creating more new trade routes, to Africa, India, the Spice Islands, and the Orient. A tiny percentage of the population—heads of state, peers, merchants, and bankers—invested and reaped this speculative wealth. In small companies, they outfitted voyages of exploration believed to promise future profit. Columbus, Cabot, Verrazzano, Frobisher, Davis, and Hudson all found a part of the Americas by seeking a new route to China that would facilitate trade by lowering freight costs and hazards.

As the sixteenth century progressed, monarchs or investment groups sponsored rare voyages to scout out land for colonization—Cartier in 1535, Gilbert in 1583, and Raleigh in 1585. Farther south, heavily laden Spanish treasure ships regularly harvested the treasures of South and Central America via the Caribbean ports of the Spanish Main, and Spanish outposts dotted Florida and California before any British colonization of North America. All the while, the fishing boats roiled the waters of the Grand Banks and sometimes Georges Bank. By the late sixteenth century, fur traders were passing them, en route to Canada and New England.

All the while, the capital of investors nudged forward exploratory searches for new ways to extract wealth from the new lands. Queen Elizabeth, for example, in the process of trying to figure out how to exploit England's land claims in the New World, supported the voyages of Gilbert and Raleigh, and also participated with a group of investors backing Sir Francis Drake's looting of Spanish ships and ports.

Finally, in 1606, commercial interest in the New World intensified to the point that a group of aristocrats and wealthy merchants organized into a "joint stock company,"[3] to whom King James, eager to harvest the fruit of his distant landholdings, gave vast, overlapping territories and expansive rights. To the Virginia Company of London he gave all of North America from approximately present-day Atlanta to Manhattan. To the Plymouth Company he gave a territory spanning from roughly modern Philadelphia to Canada's Maritime Provinces. Ownership of the overlapping portion of the grants, a strip of Mid-Atlantic America, would be settled by the race to colonize. With the French exploring what is now Canada and Maine, King James urgently needed to promote colonization to hold his claims. He hoped for mineral wealth like that found in Spain, even though it had eluded Elizabeth. The grants he made guaranteed him a fifth of any mineral wealth found.

With colonization framed as a race to realize profits from a capitalist venture, the search for gain—and for the raw materials to fuel Britain's economy—now accelerated. Transatlantic voyages increased. Along with the traffic of other nations, the Virginia Company and the Council for New England (initially the Plymouth Company and Northern Virginia Company) sponsored voyages for the purposes of exploration or colonization in North America, though their successes were spotty.

In 1607, as most American schoolchildren know, exploration and a few fruitless British attempts at colonization finally resulted in the first permanent North American colony. While that year the Northern Virginia Company sponsored a small colony that shivered through one Maine winter before bolting for home, the Virginia Company of London sponsored a group of men dying in droves on the malarial coast of Virginia. (Annual death rates in Jamestown hovered around 50 percent for the first decade.) Within about a decade, which saw the near financial ruin of their sponsors, these colonists learned enough arts of survival to allow them to plant and profit somewhat from tobacco. And less than a decade after that, in 1624, King James revoked the company's charter and claimed the colony for himself.

From this early moment in the seventeenth century, the transatlantic traffic from Britain would multiply as colonization found greater successes and the raw materials harvested by colonists found more markets. After the fish and furs of North America came timber. Then the crops of tobacco, sugar cane from the Caribbean, and later rice, indigo, and eventually cotton from the southern Atlantic colonies entered the trade routes that ran from Europe to Africa, to the Caribbean, and to the British North American colonies.

Now king and fisherman, peer and farmer, merchant, sailor, and servant all had roles in the New World economy. The New World offered resources for all, and all were needed. The upper classes provided investment capital for trade and small industries; the educated classes helped to shape initial, transitional forms of government into which freemen gradually insinuated themselves—more successfully in New England and the Mid-Atlantic states than in the South. The artisans and skilled laborers who helped to refine the new settlements were rewarded by high wages in a resource-rich, man-poor economy.

And working the land and building the first crude homes in every region were the great majority to whom "economic opportunity" meant subsistence farming: dodging starvation through continual hard labor but with nature as the taskmaster rather than an overseer, first mate, or lord. Coming from England, where rates of landownership in rural areas ranged between 10 and 20 percent of free males, immigrants to America found themselves in a world where, by the eve of the Revolution, about 70 percent of freemen owned land. Abundant land, by permitting wide distribution of real property, began the social transformation of Europeans into Americans. The wide distribution of property permitted more democratic ideals to gain hold; it mixed social classes; it gradually increased independence and privacy, weakening society's intrusiveness but also its claims to allegiance. Bountiful land supported individual welfare while the settlers' individual energies in cultivating the land supported the claims of empire.

EARLY NEW ENGLAND

For our first look at the coast of seventeenth-century America, we can have no better guide than John Smith. Samuel Eliot Morison describes Captain Smith as a lovable liar in the Elizabethan mold of good storytellers.[4] Others have called him worst. Most point out that there is no evidence to support Smith's hallmark tale of being rescued by Pocahontas. But whatever else he was, Captain John Smith was a visionary—visionary enough to see the potential story in any experience and visionary enough to recognize the potential European uses for a "New World," even one lacking gold and silver.

In 1614, the Northern Virginia Company hired Captain John Smith of Virginia fame to explore the coast of what Smith would christen "New England." Smith was to assess the land's potential for colonization, and his cruise along the coast resulted in his promotional pamphlet called "A Description of New England," in which he praised most aspects of the land and particularly Massachusetts, which he called—referring to the land of the Massachusetts Indians—"the paradise of all these parts." (Visitors to Virginia also called *it* a paradise.)

Gliding by, Smith listed the resources he saw: a coast "over-growne with most sorts of excellent good woods, for building Houses, Boats, Barks, or Ships, with an incredible abundance of most sorts of Fish, much Fowle, and sundry sorts of good Fruits for mans use." In addition to foreseeing the great fishing and ship-building industries of future

New England, he saw "Slate for tyling, smooth stone to make Furnasses and Forges for Glasse and Iron, and Iron Ore sufficient conveniently to melt in them."

He spied the Indian gardens and the healthy appearance of the Indians themselves and surmised the land's fertility. Assessing the potential for colonization, Smith couldn't promote the country enough, projecting onto it the "climate of Devon." (Smith saw it in good weather.) "Who can but aproove this a most excellent place, both for health and fertility," asked Smith.

And just as we can imagine Smith's ancient ship bearing over the coastal waters, Smith looked at the New World and imagined . . . *us*, a civilization blessed with bountiful resources. "[H]ere," Smith declared, "every man may be master and owner of his owne labour and land." This land, he predicted, would produce enough to rival any in the world; "it might equalize any of these famous Kingdomes in all commodities, pleasures, and conditions, seeing even the very edges doe naturally afford us such plenty."[5]

Smith's enthusiastic description was, in fact, a list of selling points, similar to, if more farsighted and comprehensive than, the later descriptions that settlers in the Kentucky, Ohio, Indiana, and Illinois territories would send home to tempt friends and relatives. It resembles George Washington's later description of his Ohio lands to a prospective buyer: "there can be no finer land in that or any other country; or lands more abounding in natural advantages. The whole of them are washed by the rivers I have mentioned . . . are furnished with streams . . . stored with meadow ground . . . and abound in fish and fowl." In Smith's and Washington's selling points lay not only evidence for investors but evidence of a beneficent providence and a call to those who longed for it: a huge swath of the English population, plagued by poverty and hunger. By the time of Washington's letter at the end of the eighteenth century, the Acts of Enclosure had deprived England's rural poor of even more scraps of land, while across the Atlantic, land, along with the hope of the beneficent providence it represented, was selling like hotcakes.

Smith's visionary pamphlet stirred the imaginations of numbers of his countrymen. Morison credits Captain Smith with inspiring the sponsors of the Massachusetts Bay Company and negotiating with the Plymouth Puritans, to whom he offered his services as a scout. They declined, saying that his books and maps offered a cheaper way of getting his services.[6]

However, the future inhabitants of Plymouth did not always find themselves in a strong bargaining position. The Merchant Adventurers of London, with a grant of land near the Hudson, outfitted with ships on extremely disadvantageous terms, this small group of Puritan exiles living in Leyden in Holland. Following a storm, the captain led them instead to Cape Cod, where, after some exploration, they found an abandoned Patuxet Village with remnant cornfields, which they named Plymouth. An epidemic of European germs had wiped out the native inhabitants.

As we know, the Plymouth colonists' initial trials were severe. They lost half their company that first winter of 1620–21. They needed help from the Indians and provisions from Virginia to survive through 1623. Further, two of the first ships they loaded with fur and timber to repay the Merchant Adventurers were lost. But as we also know, they survived and eventually prospered. They applied for a legitimate grant to their land from Lord Warwick of the Council for New England, which eventually granted them the area of Plymouth and some surrounding lands, including Cape Cod. Their prosperity grew largely from the continual influx of population to the region, especially to the north, where thousands migrated to the Massachusetts Bay Colony beginning in 1630. Newcomers needed supplies. Old comers produced them. Population growth fueled an expanding economy in which demand was easily met by an abundance of resources.

The wave of growth that tided the Puritans into prosperity by the 1630s, when the Massachusetts Bay Colony brought thousands of new immigrants, was not the first such wave in America. In Jamestown, planters had "discovered tobacco as a marketable staple" around 1617.[7] Jamestown was an example of a commodities-based boom that depended on the English market, although that market grew, too, notably after tobacco prices came down. Plymouth, however—which had started with a communal economy and communal landownership and had given it up in 1623 for the lack of incentive it provided individuals to work—found the profit in rapid population growth in which new markets had come and knocked on the doors of simple farmers with plenty of land.

The growing prosperity and easy availability of land brought a less puritanical attitude toward property. In 1633, three years after the

Massachusetts Bay Colony brought a thousand people to Boston, the beachhead of a great Puritan emigration from England, the governor and assistants of Plymouth passed a law stating that since certain inhabitants had helped to purchase the "land within this patent" (that is, redeem their debts to the Merchant Adventurers) and that "the said purchasers are possessed but of small portions of land, many of them meane," then it was "therefore thought meet that the said purchasers shall hold/have reserved for themselves/their heirs so much land in such place as they shall judge meet/convenient for themselves."[8]

Soon thereafter, the most prominent members of the Old Colony moved off to found new towns nearby on generous parcels that they judged meet or convenient for themselves. Miles Standish and John Alden founded Duxbury with large farms. (Miles Standish died years later with five hundred acres.) The former governor Winslow and his family moved off to Marshfield. The rush of inhabitants to new land left nearly abandoned the original village lots. The government moved to reclaim and redistribute these lots lest the town be "dispeopled."[9] So appeared the distinguished origins of American sprawl.

Governor Bradford lamented the loss of community-minded Puritan virtue to the economic engine of the New World.

> Some were still for staying together in this place, alleging men might here live if they would be content with their condition, and that it was not for want or necessity so much they removed as for the enriching of themselves.

The economic incentives that attended the Anglo-American frontier continually undid the settlement patterns of the European village—starting in Virginia and Massachusetts and proceeding across North America. When the Anglo-American frontier reached the tight-knit French town of Sainte Genevieve in the territory of present-day Illinois in the eighteenth century, "more and more frequently, innovative habitants eager to profit from the growing immigration of Americans moved their houses [away from the village] to isolated fields planted to whatever crops seemed most lucrative."[10] Not only was land more available than in Europe; it offered economic and political freedoms to a much larger percentage of the population. It was valuable as a means of production to satisfy an expanding market, as a place to live and have dominion, and as a commodity to sell to someone else.

While in Britain primogeniture had kept large landholdings in the hands of a small number of families, the economic value of land in America—to the many rather than the few—as a means to production and as a commodity—became the guiding principle in its use and the basis for a booming land market founded on speculation. In the New World, community unraveled to the tug of growth, gain, and independence, all supported by plentiful resources. Community relationships gave way to the individual relationship between a man and his land—a different economic relationship than the huddling together for survival of the earlier agricultural villages. The intoxicating lure of available land, with its broad invitation to Everyman and its promise of gain and independence, underlies the settlement of the nation and helped to shape our political institutions and public attitudes. Land, together with frontier liberty, supported equality and the individual's productivity. On the other hand, it tempted the individual to move away, to shed cumbersome social bonds, and to overlook community interests. These patterns have continued throughout American history. Now, land for housing in the large metropolitan areas where most of us live is becoming too scarce to support equality—but it continues to tempt the individual to uses that do not always benefit the community.

RICH MAN, POOR MAN

Throughout Colonial America, the land distribution process favored the wealthy insofar as it reflected entrenched European economic and social systems. New York was given in tracts to the favorites of the Duke of York after he chased the Dutch out. There, numerous privileged landlords maintained an almost feudal hold over the farming tenants on their estates that would not be disrupted until ten years after the Revolution.[11] In Virginia, Lord Fairfax ended up with five million acres, and his land agent, Robert Carter, with hundreds of thousands, an empire even by the standards of other prominent Virginia families, who gave him the nickname "King" Carter. The common man was left with the crumbs from the table, but, as we have seen, in a land of abundance, those crumbs could be sizable chunks.

In the South, notably Virginia and Maryland, land was distributed by headright. Any man who had paid his own passage received 50 acres. In some regions, during some periods, allotments of up to 150 acres per individual were made. The newcomer received an additional allotment equal to the first for any additional passages he had paid for family, apprentices, servants, or slaves. He could choose the location of his plot. No obligation to the community entered the bargain. It was understood that a man's eagerness to make his fortune would work in the colony's economic interest. This faith in development still prevails in America although, in communities overwhelmed by growth, this faith is being tested, questioned, and tempered.

For their part, the Puritan fathers tried to stem the tide of speculative activity. Land grants for towns were divided into lots of about 30 acres, which were approved by the community and then assigned, with adjustments for the individual's status. Complaints after assignment were not allowed. The Plymouth fathers insisted that newcomers to any community be approved by the proper authorities within the community and take an oath of allegiance to God, king, and colony. No one was allowed to sell or rent land to an unapproved individual. Since lots were granted for "living upon them for the maintenance/strength of society," unoccupied lands would be reclaimed by the government. Even with a grant, remote places could not be settled "without such a competent company or number of Inhabitants as the Court shall Judge meet to begin a society as may in a measure carry on things in a satisfactory way both to Civil and Religious respects."[12]

In the neighboring Massachusetts Bay Colony, town founding proceeded apace. The early history of New Plymouth—in which the town fathers held so much land that it took them many tries to give most of it away—repeated itself many times. Records from seventeenth-century Massachusetts towns such as Dedham, Sudbury, and Andover describe the same process of land division. Historian Phillip Greven described the initial process of land distribution in Andover in his book, *Four Generations.*

Andover was founded around 1645 by a group of eighteen men who were later joined by others. By 1660, the town founders had given out forty house lots in sizes ranging from four to twenty acres, according to the recipient's social status. Every man received, along with his house lot near the common, an equal number of acres of field and forest. The hay from fields saved a man the trouble of clearing land and growing food for his animals at a time when land was plentiful but labor was scarce.[13] Wood allowed him to build a home and outbuildings and the fires needed for warmth and cooking.

After distributing lots to the forty households of the community, the town fathers of Andover still had sixty square miles of land left. They distributed again and again until, by the fourth division in 1662, even the humblest members of the community, who had started with a house lot of four acres and a farm lot of four acres, received eighty additional acres, nudging their total acreage over one hundred.

As latecomers arrived to join the town and the region, the land belonging to earlier settlers increased in value. As Greven put it, "Those who arrived after 1662 were to discover why it paid to be among the pioneers. Thenceforth, the generation of first settlers would control most of the land in this community, with far-reaching consequences for their children and families." This process continues today on a more modest scale. Communities favored by growth can watch their real estate values double in a decade. The earliest residents make the biggest profits, if they care to sell their land and move.

Settlements multiplied quickly in this way, and the land cast its intoxicating spell. On the eve of the Revolution, about two hundred towns dotted New England. Throughout the colonies, about 70 percent of adult free males owned land.[14] Although the wealthy obtained more land more easily, the bounty of providence was becoming every man's right—within the limited seventeenth-century English definition of *man*. (Women and most Native Americans and African Americans would wait much longer for property rights.)

Although most Massachusetts towns had ordinances such as the one in Andover that prohibited the construction of dwellings other than sheds on land other than house lots, by 1680, half of Andover's inhabitants had moved to their outlying farm lots and the rest of Massachusetts residents were similarly situated. Social incentives to move to outlying land reinforced the economic incentives. Historians have occasionally described these social incentives in trying to counter our sentimental notions of the small Puritan village with a dose of reality, ascribing the stew of petty litigation, and even the witchcraft hysteria of

Salem, to the unhealthy effects of living at close quarters in a repressive society featuring an autocratic family structure. Later, J. B. Jackson would offer a similar motivation for those drawn to the frontier in his essay "The Westward Moving House."

> The reason given by the Tinkham tribe for the young man's defection was that there was more money to be made in farming out west. . . . But . . . when Pliny moved West it was not so much in search of easy money as in flight from the Old Testament household, the old self-sufficient economy; in a way it might be said he was fleeing the New England village: common, meeting house, and all.

Puritan beliefs could not hold forever against the promise of individual prosperity, which carried with it the whiff of individual fulfillment unrecognized in Puritan dogma. The westering towns leapfrogged inland from the Atlantic coast, and the frontier sprawled westward, propelling the evolution of a highly individualistic, self-determining spirit. Even before the Revolution, the Puritan's careful model of town planting died for good before reaching the western border of Massachusetts. In 1762, in Boston's old State House, "nine newly laid out townships in Berkshire County were auctioned off to the highest bidders."[15] The speculators who bought them had no thought of town making. They allowed the industry of the settlers to increase the value of their lands, which they generally enjoyed tax-free if they did not improve them. This disincentive ensured that the speculator would not contribute, even by paying taxes, to the improvements so badly needed by the settlers such as roads and bridges.[16]

After the French and Indian War in 1763, the British government drew, by royal proclamation, a line along the Alleghenies, prohibiting further expansion by the land-hungry colonies and dampening, at least temporarily, the sparks of any further expensive conflicts with the Indian nations of the American interior. Pioneers, settlers, and speculators crossed the line as easily as they had moved away from the pious heart of the New England village. Land companies, unable to act with the anonymity of individuals, repeatedly petitioned various branches of British government to grant them exceptions to the law and to recognize their claims to territories around the Ohio River and west of it, and they continued to argue among themselves after the Revolution. The frontier beckoned with the promise of endless growth and continual opportunity for men—and the rare woman—to enrich themselves. Nothing could check the impulse of European Americans to grasp that opportunity.

When, after the Revolution, the new federal government authored its earliest ordinances for disbursing the new federal lands (those western territories ceded by the original colonies), New England's representatives exerted their influence to see that the land was sold in town-sized chunks.[17] But this effort failed for a number of reasons.

As historian Benjamin Hibbard put it, "the [New England–style] plan did not suit the disposition of the pioneer." Instead of the Puritan plan, the southern method was adopted within two decades.[18] Under this method, the settler could choose his own plot of land, without regard to affiliating with a proposed town. This represented a surge in the contagion of speculation spreading through American culture, but Congress stepped in to control the worst damage. Initially, Congress had sold land in such large parcels that, as Jefferson's secretary of state, Albert Gallatin, reported, only speculators could afford land. Although the government revamped its land policies many times to reduce the speculator's advantage over the settler, its policies interfered minimally in frontier economics with the result that the frontier burned westward across the country, a continual wildfire of speculation. Rapid settlement spurred rapid economic growth that impressed and startled foreign visitors, as well as American observers.

By the mid-nineteenth century, Boston intellectuals such as Nathaniel Hawthorne, author of *The Scarlet Letter*, were ridiculing the hypocrisy and intolerance that had constricted Puritan society. Ralph Waldo Emerson framed the Puritans' founding contribution as a distant memory. Emerson, who traveled the Midwest and West at mid-century, boasted: "It was commerce that built this country, however Religion may have mixed on one or two conspicuous occasions therewith." To the great engine of economic growth, Emerson credited "the fruitful crop of social reforms: Peace, Liberty, Labor, Wealth, Love, Churches of the Poor; Rights of Women; all ethical—and sought not by the malignant but by the good."[19] And if Emerson enshrined enterprise, who could be expected to dethrone it?

It would be centuries before Americans looked back en masse with nostalgia at the villages of New England and the lost elements of community discarded with the Puritans' repressive strictures. Those centuries were devoted to the acquisition and settlement of a rich and promising land; to the honing of individual liberties and the building of individual prosperity; and to the attainment of every luxury that abundance could offer.

CHAPTER 3

"FOR THE ENRICHING OF THEMSELVES"

That whereas lands are given/granted to persons upon supposal of their living upon them for the maintenance/strength of society. If it fall out that persons shall not occupy any such land but depart from the same place, such former grant or grants to be of none effect but shall return/be otherwise disposed of by the Government in general or Township in particular as it shall fall out.

—"Laws," Plymouth Colony Records

The New Englander is attached to his township because it is strong and independent; he has an interest in it because he shares in its management; he loves it because he has no reason to complain of his lot; he invests his ambition and his future in it.

—Alexis de Tocqueville, *Democracy in America*

On Cape Cod, the old Puritan towns continued to govern themselves through public meetings, the kind that impressed Alexis de Tocqueville as the very cradle of democracy. You can still go, as I did a couple of times during my first planning job, to hear the melding of voices and arguments and to observe the patient parrying that makes up self-government. Even now, the voices of the old Puritan fathers echo at these meetings, arguing the community's best interests. Neither the old town meeting or the Puritan ethos, however, has proved a match for the modern clamor of the American economy and the opportunism of outsiders, large-scale construction, and rapid population growth.

As the frontier had burned westward, Cape Cod's soils had responded to the efforts of Puritan farmers by blowing away.[1] But the sea sustained a culture of fishermen and seamen who manned a packet-boat trade between New York and Boston, and longer voyages out of Boston. "The shores," Thoreau wrote of Cape Cod, "are more fertile than the dry land."[2] But not long after Thoreau wrote those words, Cape Cod lost its fishing industry to deeper harbors elsewhere, its packet trade to trains, and Boston lost shipping ventures to New York. Cape Cod's population blew away like the fragile soil, reduced by a quarter.

Born of the insular Plymouth Colony and further isolated by geography, with its old-time seafaring culture preserved by a stagnant economy, Cape Cod grew quietly exotic. Its windswept beaches saw mainly birds and occasional clam diggers. But in a nation of increasing wealth, speculative real estate investment had discovered resort areas late in the nineteenth century. By the 1920s, the resort real estate boom was in full swing and Cape Cod was ripe for the picking.

John Kenneth Galbraith describes the Florida real estate boom of this period in his book *The Great Crash 1929*. "The Florida real estate boom," Galbraith concludes, "was the first indication of the mood of the twenties and the conviction that God intended the American middle class to be rich. But that this mood survived the Florida collapse is still more remarkable. . . . Even as the Florida boom collapsed [in the late twenties], the faith of Americans in quick effortless enrichment in the stock market was becoming every day more evident."[3]

At the same time that the warm Florida climate cast its spell, during the 1920s, the Cape's scenic harbors and beaches, her villages of white clapboard homes with lilacs at the corners, and other antiquated charms were beginning to appear in the pages of national magazines as a little-known tourist dream. The nation that read these magazines was becoming thickly settled. The nation's arable public lands had been claimed by homesteaders or speculators. Major American cities had long since grown into large metropolitan areas, riddled by industry and slums, spawning both suburbs and an appreciation of vacation resorts, both initially for the affluent.

The first articles about Cape Cod described its unique natural setting, its historic villages and Yankee culture, its cranberry farms, and the artists' colony at Provincetown. Henry Kittredge, who wrote some of those articles, described the effect of the new trickle of tourists in his 1930 book *Cape Cod: Its People and Their History.* "The Cape Codder feels no qualms about nailing up roadside apple stands for the automobilists, or filling his spare rooms with transients for the night. Some have built shore-dinner restaurants or converted the old barn into an antique shop for the summer visitors."[4]

The response of Cape Cod residents to an influx of newcomers was not so different from the response of their Plymouth ancestors three hundred years before. They saw an opportunity to which they applied their efforts with gratifying results. However, Cape Codders were not the only ones taking advantage of the growth. Real estate speculators and developers had already begun to circle by this time. Speculative land purchase and development, honed in the West and in the development of suburban subdivisions around the country, was now as American as apple pie. Gone were the Puritan ethics that once governed land distribution on Cape Cod. And as visitors and investors recognized the advantages of Cape Cod, the land rush gathered momentum.

On Cape Cod, the land rush proceeded more slowly than the Florida rush of the same period—and without the ensuing bust. With their ancient, charming town centers framed against the ocean and their relative proximity to Boston, the towns of the Cape became tourist destinations.

The Cape Cod towns hit by early growth responded with the new tool of zoning to channel growth. As homes crowded onto the neat swatches of single-family residential districts on early zoning plans, however, towns lost scenery and open space and discovered the limits of zoning. These recent historic developments on the Cape—population growth, real estate speculation, and the response of communities trying to control growth—would eventually produce a particularly tense public meeting that I witnessed on Cape Cod, just as they produce similar standoffs around the country. Because I had come to that meeting as an observer, and not to persuade or represent anyone, I was free to listen to the voices of Cape Cod residents anguishing over the impact of prospective development without trying to offer quick answers. I was free to become a writer instead of a planner, to stand farther back, and to puzzle over the interests of those speaking and listening.

One warm late afternoon in 1988, my soon-to-be husband and I got in the car to drive from Boston to a town on Cape Cod that I had never heard of. The town of Mashpee had somehow occupied an interior portion of the Cape for centuries without attracting much notice. Its southern border was Atlantic waterfront, but smaller than the waterfront of any other Cape Cod town. At first, the visitor might guess that Mashpee's interior location has worked against it, since all the life of Cape Cod seems to occur at the water's edge, and this littoral lack will not have helped to promote the town. However, a little digging reveals Mashpee's unusual origin, in 1654, as a "district for praying Indians." Mashpee was finally incorporated as a town in 1870—over two centuries after its own founding and the founding of the Cape's oldest towns. At this point, the lands of Mashpee were formally opened to the booming American real estate market, and a few decades after that, the seeds of the 1988 meeting we attended were planted by one Malcom Chace.

Malcom Chace, the son of a Rhode Island industrial magnate, looked appreciatively on the scenery of Cape Cod. According to records in the Barnstable County Registry of Deeds, Malcolm Chace, acting as a party to Nantucket Sound Associates, purchased several large shorefront parcels in Mashpee in January 1929.

The Depression that followed the October stock market crash that year delayed development on Cape Cod. At first, summer camps and modest seaside cottages sprang up along the ocean. Malcom Chace opened his property to seasonal camping on a fee basis. With economic recovery, however, came more substantial development around the Cape. Chace began building on Daniels Island, reached by causeway from Popponessett Beach. The properties' manager wrote "To Former Campers at Popponessett" to announce the management's plan "to erect each year a limited number of summer cottages."[5]

This twentieth-century tide of growth differed from the seventeenth-century tide that brought the first towns to Cape Cod. Land and resources were now scarcer on Cape Cod. After a few decades of twentieth-century-paced growth, there were very few, very costly lots on

Cape Cod to which people could move. There was no land on which to build new towns nearby the old ones as the Puritans had.

In Mashpee, Malcolm Chace's purchase of Mashpee land, followed by further family acquisitions in the next generation, would affect Mashpee throughout the twentieth century through developments such as Popponessett, built in the 1940s and hailed as a model of sensitive resort design; New Seabury, begun in the 1970s with an array of time-sharing condominium units and still growing; and Mashpee Commons, first undertaken in the 1980s and, though projected to have thousands of units, taking shape very slowly to date.

In the late 1940s, vacation-related growth accelerated on Cape Cod, blending new money with the old seafaring culture. By the 1950s, the tourist, resort, and leisure boating industries had brought growth back to the Cape with a vengeance. Throughout the second half of the twentieth century, the towns of Cape Cod and the nearby islands of Nantucket and Martha's Vineyard were seeing annual growth rates of 60 percent or more. Between 1940 and 1960, the entire population of Cape Cod doubled.

The 1980s brought a building boom to all of eastern Massachusetts, including Cape Cod, and then a brief recession. After years of booming growth, the bulldozers on Cape Cod were still. Cape Codders, as they periodically had, saw the other side of their troubling conundrum: too much growth—especially of the wrong kind—and the delicate beauty that drew tourists and retirees could erode like the fragile soil; too little growth and unemployment and business declines crept in. However, we must distinguish between the local economy and the regional or national economy here. The local bed-and-breakfast owners' losses differed from the outside speculators' who invested in large construction projects. Managing growth to benefit the local economy and channeling growth into environmentally friendly industries still seem to hold the greatest promise for places like the Cape.

In 1998, with the boom again in full swing, *USA Today* declared the Cape and the islands to be part of a "Coastal boom" affecting one hundred "hot" counties along the Atlantic and Gulf coasts. The article declared that "more than one in seven Americans now live in a county that abuts the eastern or southern seaboard. That number swells by several million when inland residents with second homes near the shore are included."

On Cape Cod, commuters to Boston looking for bedroom communities had by then crossed the Cape Cod Canal. In the towns nearest the mainland, the unique natural beauty of the sandy peninsula had given way to heavy residential development and tourist schlock. By 2000, Cape Cod's population had reached 222,230—more than the population of all thirteen colonies combined in 1690, around the time when the governments of Plymouth and Massachusetts Bay merged unhappily under the king's royal governor and several years after Mashpee's founder, Richard Bourne, pastor to the early Indians there, died.

When Alex and I arrived in Mashpee, over a hundred years after Mashpee's incorporation, we were looking for a public meeting concerning a real estate development proposed by Malcom Chace's grandson. As we turned onto a thinly settled road in the waning light of late afternoon, we passed the town hall, sitting almost alone at the intersection. After passing through a couple of miles of woods, we found more buildings: a small, rather lonely shopping center at a large highway rotary, where we parked in the last, long shadows of the day and found the meeting that was about to begin.

Inside, a handsome young Rhode Island developer with the mien of a yachtsman, Buff Chace, Malcolm's grandson, was to have his design team present a plan for the shopping center and some acres around it—a plan to turn it into what Chace and his designers called "the town center that Mashpee never had." Chace had tried to adhere to the high family standards that had earned awards for the family's earlier developments. And yet, the sands of Cape Cod had shifted. The meeting was taking place only a few months before a majority of Cape Codders would vote to establish the Cape Cod Commission to help stave off large, new developments being proposed around the Cape.

Buff Chace had hired some of the most thoughtful, idealistic designers available. Andres Duany and Elizabeth Plater-Zyberk were soon to be celebrated as among the chief architects of the "New Urbanism," a scheme to reshape the suburbs by restructuring subdivisions to resemble small towns, to restore classic principles of town planning to the suburbs. Their high standards in designing residential communities were appreciated almost everywhere they went.

However, when the slides of a beautiful, simulacrum New England town faded and Andres Duany opened the floor to questions, the residents of Mashpee sounded testy and skeptical. The potential impacts of the project troubled them. What about traffic congestion? What about water quality? What about the bass in the Quashnet River?

The Mashpee residents who attended the meeting that evening were not interested in development, no matter how virtuously clothed. They were interested in the quality of life they had settled or stayed on Cape Cod for and in the fragile ecosystems of the beautiful scenery that supported that quality of life. And behind all their questions was a greater one. No matter how many times Duany explained that the property was already zoned commercial and that the town could be faced with two cinemaplexes instead of their thoughtful community design, the townsfolk did not understand how this could happen without their consent.

The townsfolk were really asking why the developer's right to make money from his land outweighed their right to protect their way of life and the natural and scenic assets that they had bought into and that were being slowly eroded by development. The community-minded principles of the Puritans echoed around the room. And so did headlines from around the nation, in which more and more Americans questioned additional development and the continual trumping of community interests by profit. Could growing population in America be curtailing the long reign of individual interests? It is only what seems to be the beginning of a trend in areas of prolonged rapid growth. The property rights movement has answered loudly, and we know there will be many fronts in this battle.

The Puritans knew what to do when community interests were threatened. Except for a blind spot where Native Americans were concerned, they believed in the principle of first come, first served. When, in 1636, they observed the worn-out condition of the ground around their fishing stream, they passed a law:

> That whereas God by his providence hath cast the fish [called] alewives or herring in the midst of the town of new Plymouth. And that the grownd thereabout hath been worn out by the

> whole to the damage of the those that now inhabit the same. It is therefore enacted That the said herrings Alewive or shadd commonly used in the setting of Corn be appropriated to such as doe or shall inhabite the Town of Plymouth aforesaid. And that no other have any right or propriety in the same save only for bait for fishing, and that by such an orderly course as shall be thought meet by the Governor Assistant.

Though impressive, the Puritan model depended on an abundant land supply. If a town had too many residents, they started a new one. When, in 1668, several towns appeared to have *too few* residents, Plymouth declared a moratorium on new towns for seven years until existing towns could be populated to a satisfactory degree.

Nearly four hundred years after the Puritans wrote their laws, Cape Cod had run short on land, short on housing, and American history had strengthened the concept of private gain and speculative profit as an American right. Still, in the questions of irritable Mashpee residents at that evening meeting, a faint echo of the Puritan concern with the health and interests of the community sounded. But how could residents advance those interests?

The townspeople were at the mercy of a system of land use planning and regulation that had become so complex and economically weighted as to outgrow the New England town meeting that de Tocqueville admired. Though this was not an official town meeting, these townspeople were frustrated by being assembled for what seemed to be an exercise of their power, only to find that they had no power. The future Mashpee already existed in a large report full of fine print and diagrams in the town planner's office, and the townsfolk seemed unaware that that future Mashpee contained many more stores, homes, and people than the one they inhabited. By 2000, the woods we had driven by would all be purchased and platted for subdivisions. And this despite conscientious conservation efforts on the part of the town's planner, conservationists, and others to save fragments of open space and environmentally sensitive land—around town wells, for example. This future Mashpee evolved out of the continual pressure of increasing population and economic growth and because of the inadequacy of tools such as zoning to preserve land in the face of development pressures and the inability of other planning tools to save land by increasing housing density in the face of consumer preference for lower densities. It was not helped by a local zoning appeals board that rubber-stamped most applications.

Cape Cod residents had become so frustrated by these defeats that their representatives were trying to formulate strategies for halting or managing growth on the Cape. Months after the meeting I witnessed, Cape Codders voted to surrender some of the cherished local autonomy embodied in their town meetings in order to create a planning authority with regional jurisdiction.

The founding of the Cape Cod Commission, finally established in 1990; the earlier questions of the Mashpee residents concerning new developments in their town; the establishment of tiny real estate taxes on neighboring Nantucket and Martha's Vineyard at about this time for the purchase of public land—all these developments showed changing American attitudes toward growth in the twentieth century. In that century, the U.S. population more than tripled from about 85 million to about 281 million and modern technologies facilitated the rapid transformation of the landscape.

Our changing attitudes are to be found throughout the landscape, throughout the fine print of court decisions and zoning laws, and in the heated debates in public meetings around the country. For in modern America, it's not just the landscape that is changing; it's the American.

CHAPTER 4

HOME IS WHERE THE REGULATIONS ARE

Is there not the earth itself, its forests and waters, and all other natural riches, above and below the surface? These are the inheritance of the human race, and there must be regulations for the common enjoyment of it. What rights, and under what conditions, a person shall be allowed to exercise over any portion of this common inheritance, cannot be left undecided.
—John Stuart Mill, *Principles of Political Economy*

Family legend has it that Archibald Funston, my paternal great-great-grandfather, lost the family lands on the Firth of Tay to the rule of primogeniture around the middle of the nineteenth century. He left the land where his ancestors had been marrying Crawfords and Campbells for generations and spent some years in Northern Ireland. There his wife gave birth to their daughter, Elizabeth, in the town of Trillick in 1855. When Elizabeth was about twelve, the story goes, the family sailed to America, where they bought land near the Hudson River in the suburbanizing rural community of Brooklyn. On that land, young Elizabeth grew up. When she married, she moved uphill to Bay Ridge. There my grandmother, Isabel Fuller, was born and grew up. And there, in 1920, my father was born, about a year after my grandfather returned from World War I.

In that time between my great-great-grandfather's arrival around 1870 and my father's birth, the populations of New York City and

Brooklyn, receiving the brunt of immigration to America, had both quintupled. New York City had grown by about a million people per decade until, in 1920, the city held over five and a half million people. Brooklyn's population had grown from four hundred thousand to about two million people. Ferry routes, the construction of the Brooklyn Bridge in 1883, and train lines had all knitted Brooklyn to the city.

And although no other American city could boast New York's population, the rest of the country was also experiencing rapid urban growth. This new growth created new conflicts as people crowded together in cities, built suburban homes in farming neighborhoods, and watched new types of industry, power generation, and transportation flourish in their midst and transform their neighborhoods. These conflicts occasioned first a growing body of lawsuits among property owners and then new kinds of land use regulation and new ideas in town planning that attempted to stem both the conflicts and the lawsuits. A new concept called "the building zone" had arrived—that is, a zone for which local statutes dictate the use, size, and placement of buildings.

Experimentally in 1923, and then officially in 1924, the U.S. Department of Commerce issued the work of a committee that included Herbert Hoover and Frederick Law Olmsted Jr.: the Standard State Zoning Enabling Act: Under Which Municipalities Can Adopt Zoning Regulations. In August 1924, the Department issued a press release declaring that thirteen states, in many different regions, had adopted the Standard Act and that the act had either been adopted or was "being considered by nearly every state in the Union." New York had already enacted zoning enabling legislation, as had New Jersey, responding to pressures on its northern borders from the impinging expansion of New York City.

The newness of zoning laws made them suspect, but in 1926, the Supreme Court accepted an opportunity to officially bless zoning. Somewhat earlier, a small Ohio town near Cleveland, the town of Euclid, had established a zoning plan in an effort to stave off the industrial growth that was sprawling out from Cleveland. In doing so, it created residential and business zones where industry could not be introduced, thereby disappointing those who had bought up real estate near the railroad tracks in the expectation of profits from industrial growth in Euclid. The ire of the Ambler Realty Company, which had experienced such a disappointment, propelled the case of *Euclid v. Ambler Co.* all the way to the Supreme Court. In addressing the quandary of the town of Euclid, Justice Sutherland wrote for the majority in favor of Euclid's right to stem industrial development and against the suit brought by the Ambler Realty Company:

> Building zones are of modern origin. They began in this country about 25 years ago. Until recent years, urban life was comparatively simple; but with the great increase and concentration of population, problems have developed, and constantly are developing, which require, and will continue to require, additional restrictions in respect of the use and occupation of private lands in urban communities.

The continent was settled. Urban growth continued apace. And finally America was crowded enough to bring into collision the ever more popular habit of real estate speculation with the ancient concept of community protection and the rights of the homesteader and homeowner so cherished by Americans and their land policy. Advocates of property rights often portray land use regulation as a curtailment of property rights, but these regulations more accurately mediate between the rights of different claimants, with the government sometimes representing the interests of the majority or the community.

In the Euclid decision, Justice Sutherland went on to enumerate the threats to the homeowner's welfare represented by modern urban life. In zones restricted to residential development, Sutherland explained, homeowners and their children would be safe from the traffic, crime, and loitering associated with places of business, which "are noisy; . . . apt to be disturbing at night; . . . malodorous; some are unslightly; . . . apt to breed rats, mice, roaches, flies, ants, etc." These protective principles—expressing the zeitgeist that fueled suburbanization throughout the twentieth century—made zoning an issue of public health and safety. After all, an organized building pattern allowed for greater efficiency of fire and police protection as well as the protections Sutherland mentioned. Zoning was allowed as a proper use of the state's police powers to guard that public health and safety.

Although Justice Sutherland noted the societal growth and transformation that led to zoning, his decision invoked the historic federal sympathy for the homeowner, championing the homeowner in the face of commercial speculation. Indeed, Sutherland's desire to safeguard the homeowner against traffic, noise, and loitering strangers seems almost romantic. But Sutherland was protecting homeowners from business and industry. By the mid-twentieth century, homeowners began to ask for protection from each other—from the deleterious effects of added population and the commerce and traffic that followed it as surely as night follows day. That protection has been harder to come by.

While zoning spread around the country, and houses around the Fuller property grew closer together, young Grandma Isabel took her boys—my father and uncle—and followed her husband to a series of army bases. While their father was promoted through the ranks to colonel and given different assignments, my father and uncle got rashes from mango leaves in Panama and ate turtle soup in the Phillippines. And they attained manhood just as World War II loomed.

At the war's end, my father, a reluctant captain of twenty-five who was miserable in the military culture, resigned his commission. After a few aimless years, he took advantage of the GI bill to finish college and get his master's degree.

Although my father hated military life, he couldn't seem to roll his apple too far from the tree. The global convulsions of World War II and a childhood spent partly abroad, plus a tyrannical military father, all influenced my father's career choice. He studied geography with a specialty in industrial transportation systems. This enabled him to take a government job assessing the industrial capacities of nations, which, as he later explained to me, is a nation's capacity to wage war. And so, in 1958, my mother, father, three-year-old self, and two-year-old brother were on our way to Arlington, Virginia. In that Washington suburb my father would analyze intelligence at a Defense Department installation, and my brother and I would grow up in the company of thousands of other families who had moved to the Washington area for federal jobs.

If foreign policy was my father's passion and the focus and cultural context of his generation, suburban expansion was the context of mine. Ours was a generation well insulated in its search for prosperity, coming of age at the end of the Vietnam War and not to see another war for decades.

There was hardly a better seat from which to watch the late-twentieth-century suburban expansion in America than the Washington metropolitan area. Arlington is in the first ring of suburbs around Washington, separated only by the Potomac River from Georgetown and the Monumental Core of the Capital. Our small county contained Fort Myer, a military base; Arlington Hall, the Defense Department complex where my father worked; and the Pentagon and the Pentagon's Naval Annex, where my mother would program computers and analyze computer systems after my parents divorced around 1960. (In divorce, as in their choice of the suburbs, they were again at the leading edge, or fraying edge, of a social trend.)

Though Arlington, like Brooklyn, lay across the river from a steadily growing city and was another chapter in the growth of the American metropolis, the similarities ended there. The gradual growth of the federal government in response to the Depression, war, and national growth fueled the slower growth of Washington's metropolitan area. Twentieth-century suburbanization differed mightily from nineteenth-century patterns, thanks mainly to the car, and Arlington was still a young, green suburb when we moved there—a small suburb. But my family watched some of its leafiness give way to more commercial development and apartment complexes as it matured.

From 1930 to 1950, before we arrived, Arlington's population had more than doubled in each decade, growing from 26,000 to 135,000. For the next two decades, it added about 30,000 people a year, bringing us to the year 1970, when I started high school and began to notice the small townhouse developments being shoehorned into available open spaces. Arlington was only fourteen square miles in area. There was no more room for new subdivisions of detached single-family homes. But no matter; suburbanites desiring a more rural setting or cheaper land and housing had discovered Fairfax County—lying just beyond Arlington—a few decades before. Arlington's pattern of growth was repeated in Fairfax, but about a decade later, like an echo.

Unlike Arlington, Fairfax was a large county of about four hundred square miles. By 1950, Fairfax County's population had mushroomed to a hundred thousand, and it was still growing exponentially. The eastern portion of the county now accommodated a fairly dense community—largely made up of federal workers—while the western two-thirds of the county, bordering the ancient fox-hunting grounds and Blue Ridge foothills of Loudoun County, remained in agricultural uses and rural scenery. These circumstances—once again circumstances of rapid growth as in the case of Cleveland sprawling out toward Euclid—created a new test for zoning and land use regulation.

In 1956, the year my brother was born, Fairfax was declared the fastest-growing county in the United States.[1] That year, Fairfax revised its earlier zoning plan in consideration of the county's rapid growth and the skyrocketing costs of accommodating that growth with county services of all kinds: sewerage, water supply, fire and police protection, sanitation services, schools, libraries, and administrative services. Faced with these costs and, perhaps, the vanishing rural character of the western portion of the county, the zoning board changed the zoning of the western two-thirds of the county from agriculture, with half-acre lots permitted, to agriculture with residential development on two-acre lots.

As I would learn later in graduate school, a group of developers who owned six thousand acres of land in western Fairfax among them promptly sued the county, charging exclusionary, or "snob," zoning that would limit development and population in the western portion of the county. Another, larger group of landowners supported the county. They made known their own interests in maintaining Fairfax's character and avoiding the problems of rising county debt and diminishing water supply that the supervisors foresaw.

In court, the county's representatives protested that the costs of keeping pace with growth between 1947 and 1957 had caused the county's debt to rise from eight hundred thousand dollars to almost *fifty million dollars*, as the county supplied thousands of newcomers with roads, water, sewerage, fire and police protection, services, administration, and schools without a sufficient tax base. Fairfax's debt was expected to rise again by another seventeen million dollars in the next year or two, since the bonds had already been approved. The county's bond rating had dropped because of its massive debt, causing it to pay higher interest rates on bonds. In addition, representatives for the county argued that denser development would exhaust local groundwater supplies, threatening the public health. The supervisors also claimed to have protected farmers in the area through the zoning ordinance. And although they did not say so, because the argument had no legal merit at that time, it is generally understood that the Fairfax County supervisors hoped to preserve some of the county's rural scenery and green open space in the large-lot residential zones.

None of these arguments swayed the court. The court said that agriculture was impractical in the area anyway because of high labor rates, that if Fairfax ran out of water, local officials could purchase it. The court found two-acre lots exclusionary. The court pointed to the previously platted, small-lot subdivisions in western Fairfax that the supervisors had grandfathered into the new zoning provisions. The court said the grandfather clause contradicted the urgency that the supervisors expressed. The court used this technicality to abrogate Fairfax's new zoning.

The court also expressed sympathy for the newcomers to Fairfax—developers and residents—with which it waved away the county's argument for fiscal responsibility and resource conservation. The case file summarized: "While the cost of supplying government services should be considered in determining the reasonableness of a zoning ordinance, a barrier may not for reasons of governmental economy be set up against the natural influx of citizens who desire to live in an area." In other words, no reasonable argument to limit growth had been found.

Between the 1959 court decision and 2003, Fairfax's population soared from a quarter of a million to almost one million people—a population larger than that of several states. The county's debt soared and its bond rating dropped. It turned to the Potomac River for water. By 2000, it was taking more Potomac River water than any other county in Virginia, helping to provoke a lawsuit from neighboring Maryland, which also relies on the river and claims historic rights to it.

The Virginia Supreme Court's decision placed the law's greatest protection on the economic forces of growth. Although cloaked in the virtue of hospitably accepting newcomers, the decision was seen by those familiar with the court as a prodevelopment decision. The decision ushered in two different trends. First, in sounding a battle cry against snob zoning, the court deflected attention to the issue of housing opportunity that would be taken up by courts around the nation through the twenty-first century, with mixed results. Second, the court's decision delivered a message to municipal governments around the nation about the limitations of zoning, although that message was muffled and slow to register. The decision suggested that, if the character of a town and its allotment of open space depended upon the character of privately owned land in the town, then zoning could do nothing to prevent that character from being built out of existence.

THE MYTH OF AFFORDABLE HOUSING IN THE EXURBS

For the moment, regarding the first trend—court efforts to protect housing opportunity—we should note that courts eventually recognized that, while preventing large lots might appear to continue the historic egalitarian concern of American government with the homeowner, small lots in no way guaranteed affordable housing. Small lots could be—and were—used simply to allow more people to buy expensive houses.

That is what has happened in the Virginia suburbs of Washington. That is how, in 2002, Fairfax, a county of small lots and condominiums, ended up with an average home price eleven thousand dollars higher than the average home price in neighboring Loudoun County, a place still partly rural and still with larger-lot zoning in most of the county. A recent study found land around Washington, D.C., to be disappearing under development at a rate of about twenty-five thousand acres a year;[2] simultaneously, housing prices in all suburban counties around Washington continue to rise as they have been for decades under the pressure of a continual influx of population.

In some states, courts began to demand affordable housing rather than small lots. The first and best known of these decisions was made in the New Jersey Supreme Court concerning the discriminatory housing practices of a town called Mount Laurel.

A shortage of housing does, of course, cause housing prices to rise. But so does the shortage of available land caused by development. While it makes sense to increase housing density to answer a housing shortage, the density—especially if any land is to be preserved for the recreational needs of a community—must be greater than that found in traditional single-family home subdivisions, no matter how small the lots. However, developers of these subdivisions still use the cry of housing opportunity to fight against all kinds of restrictions on development, as happened in Loudoun, where developers testifying against building restrictions at public hearings in 2002 invoked the need for housing.

The limitations of zoning revealed by the 1959 Fairfax decision percolated more slowly through the national consciousness. It was already understood that land could never be zoned for open space because that would deprive the landowner of the value of his or her land. But the news that zoning was powerless to preserve town character through lot sizes revealed the vulnerability of American communities' vast quantities of land—soon to be carpeted with development.

Zoning, however, could at least be modified, and that is what happened over the ensuing decades. Zoning became more sophisticated, framing a negotiating process between developer and community. For example, in certain zones, the community could require a particular ratio of open space per built area, or the developer could agree to sparser development on one parcel in exchange for the right to develop another more intensively. The bargaining process can be quite frustrating for developers and can still leave community residents feeling cheated by rising taxes that pay for the infrastructure or schools that support new development.

At the dawn of the twenty-first century, zoning regulations have become far more numerous and complex than the sample zoning enabling language issued by the Department of Commerce in pamphlets of only about ten pages in the 1920s. Mashpee's "Comprehensive Plan" for growth and zoning runs to nearly four hundred pages. Loudoun County's proposed *revisions* to the 1993 zoning ordinances run into the hundreds of pages.

The dawning understanding of the extent of open space to be lost led municipalities, states, and nonprofit groups to begin to buy more land to preserve for the public—for its scenic, historic, cultural, or environmental value or its importance as a natural resource. In the growing communities where open land is most needed, however, it tends to be alarmingly expensive. Even finding a lot for a new school or fire station for a rapidly growing community can be challenging—further raising the cost of growth to the public.

Planners were not the only group to attempt reforms. Developers, faced with more people and less space in the Capital region, experimented as well. In Maryland, where the Washington suburbs began to merge with the suburbs of Baltimore, the Rouse Company built Columbia, Maryland, in 1968. Columbia was built as a model community with a target population of one hundred thousand, roughly halfway between Baltimore and Washington. Its designers intended to demonstrate the possibility of concentrating growth in selected areas to avoid sprawl. Unfortunately, copycat Columbias gradually carpeted the Baltimore-Washington corridor, along with commercial strip development catering to the residents of these vast subdivisions (much as copycats of Seaside—a New Urbanist model—later carpeted part of the Florida panhandle surrounding the original). When my brother and I drove through the area to visit my father after he remarried and moved to suburban Maryland, we often became lost in the sprawling sameness.

In the western part of Maryland, suburban growth advanced more slowly upon the rolling pastoral landscapes that became known as the exurbs—until the 1980s and 1990s, when demand finally boosted the rate of growth there. In Virginia, suburban growth spreading out from Washington pressed beyond Arlington and Fairfax to the pristine lands under the shadows of the Blue Ridge Mountains.

Growth around Washington, however, was transforming not only the countryside but also the city and inner suburbs. Not only had residential development in Arlington changed from single-family homes to townhouses, but Arlington's home prices and property taxes continued to rise. Cheaper housing—or at least more house for the money—was by then a couple of counties farther out in rural areas such as Loudoun and was migrating ever southward and westward.

Traffic congestion throughout the area engendered continual highway and road planning. To reach the growing exurbs, transportation planners tried for years to popularize the idea of a new superhighway. Route 66 would cut through Arlington and Fairfax to make Washington accessible to commuters who were gradually beginning to move to outlying counties such as Prince William and Fauquier Counties, focusing growth pressures there. (The access road to Dulles Airport, built in the 1960s, had already made Loudoun County more accessible.) By my senior year in high school, Route 66 had been approved and funded, routed mainly through lower-income neighborhoods, including one where my friend Rita lived. Rita's family occupied a very small house near our high school, where they lived on her disabled father's Social Security. Rita's family's property was purchased by eminent domain. I was reminded of Rita when I read that, in November 2002, a twelve-year-old girl cried before the Arlington County Board as she explained to board members that proposed redevelopment meant demolishing the $575-a-month apartment near the Metrorail line where she and her family lived. The development proposed to replace her apartment and others with luxury apartments.[3]

Growth in the outer suburbs displaces residents of the inner suburbs in all these ways—and through rising property taxes, to which the elderly are especially vulnerable—while we celebrate the myth of cheap housing on exurban land. We always subsidize that growth—publicly and privately.

Engineers threaded Route 66 through the suburbs far more carefully than they laid highways in southeast Washington—the low-income Anacostia neighborhood. Anacostia offers a far more tragic example of the inevitable chain reaction that accompanies suburban expansion. Since the 1940s, this area has been riddled with successive highway projects built to serve the multiplying suburbanites beyond its borders. These highways have reduced local air quality, displaced poor residents, destroyed and separated neighborhoods, cut off local access to the Anacostia River and its parklands, and created large no-man's-lands devoid of economic or urban vitality.

One man's American Dream is another man's nightmare. This is the cycle of never-ending growth in America. As we multiply, the unintended effects of growth increase and multiply. Outside the city, on the exurban frontier, the lands to which the residents of the inner suburbs used to escape for scenery and recreation are being subdivided into house lots. In the Washington metropolitan area, a million new homes have sprung up since 1970 to accommodate about two million people. And as growth has continued, a greater awareness of the repercussions of growth has inflamed the debate on growth and property rights there as bordering in locales across the country more and more strident.

Loudoun County, Virginia, neighboring Fairfax, has hosted a particularly virulent form of that debate. The headlines about Loudoun began to reach me after I had left home. After I had studied the 1959 Virginia court decision on Fairfax—which had been handed down when I was four years old—in graduate school. After I had become a middle-aged planner with children.

The country too had changed, although the fight in Loudoun brought up many of the same issues that had roiled Fairfax forty years before. The similarities to the Fairfax case made observers wonder: what would win out, the conservative political forces of the Old Dominion or the changing attitudes toward environmental change that had swept the nation? After all, in 1969, the Cuyahoga River in Cleveland burst into five-story flames—again. In 1980, the toxins at Love Canal came calling on average American homeowners and caused the relocation of around one thousand families. Evidence of environmental degradation, coupled with advancing scientific understanding of human impacts on the environment, both contributed to the mood that spawned a dozen major acts of national legislation to protect America's air, land, and waters.[4] In addition, scores of smaller laws expanded the area of the National Wilderness Preservation System through the addition of individual parcels. And, finally, private conservation groups multiplied, all attesting to new national feelings of urgency concerning environmental quality and loss of natural resources as well as increasing suspicion concerning the deleterious effects of growth.

And so, as subdivisions replaced farm fields throughout eastern Loudoun, as Loudoun residents and developers from both local and national firms and lobbying groups battled, and as the Loudoun supervisors prepared to pass new zoning laws creating twenty- and fifty-acre lots in the county's western portions, observers such as I had to wonder how the county would fare. How would the exasperation of residents affected by growth—so typical of other metropolitan areas of the nation—counter the economic interests and the forces of population growth that pressed in on them?

CHAPTER 5

"ANOTHER CIVIL WAR"

Property is vigilant, active, sleepless; if ever it seems to slumber be sure that one eye is open.

—William Ewart Gladstone

Whenever I read about Loudoun County, I remembered the country pilgrimages of my childhood and adolescence—pilgrimages to our rented cottage in the mountains, to the Waterford Fair, to a rare, all-day picnic with friends at a state park, or to the house of a friend of my mother's where I caught tadpoles in the creek and found baby mice nested in a drawer. Those excursions took us to or through Loudoun County.

In those days, the drive through Loudoun was long and lovely and boring—sort of the way I imagined heaven at that age. We called out the cows and horses to each other as we passed mile after mile of rolling pasture and an occasional stone house or barn. We knew we had entered not only another place but another time. The old houses, barns, or mills of indigenous stone bespoke an earlier era of patient, solid building using locally harvested materials. The presence of cows or horses who had nothing better to do all day than to stroll and munch the beautiful countryside suggested a pace of life unimaginably slower than the one we led in Arlington—and a place slower to change its ancient impressions on the landscape. The small groupings of old houses also spoke of lost ways. I had no idea so few people could make a community.

At that time, Loudoun County was still known for fox hunting, a holdover from the colonial aristocrats who initially owned it. It was hard to believe, when I heard of the development pressures mounting in Loudoun, that its rolling pastures were falling to subdivisions. But another holdover from the colonial aristocracy was at work in Loudoun. While the Puritans were carefully regulating the growth of their towns and the tenor of their communities, members of Virginia's leading families were busy acquiring land and using political connections to do so. Indeed, Loudoun County had once been part of a five-million-acre royal grant to Lord Fairfax, whose parcel was referred to as the "Northern Neck" of Virginia. Lord Fairfax sold and rented plots of this land. His land agent, Robert "King" Carter, acquired three hundred thousand acres of prime Virginia land in the course of his work. Even a young protégé who surveyed some of Lord Fairfax's lands for sale or rental—the adolescent George Washington—used the opportunity to acquire some choice parcels as well.

Among the tens of thousands of acres George Washington acquired throughout his life—in Virginia, Maryland, Pennsylvania, New York, and near and along the Ohio River—Washington, in his will, describes his three hundred acres in Loudoun "on the great road from the City of Washington, Alexandria and George Town, to Leesburgh & Winchester" as being "valuable, more for its situation than the quality of its soil." This well-located parcel, he noted, fetched considerably more per acre than did the more remote three thousand plus acres he owned that straddled the Loudoun-Fauquier county line—more per acre than any other piece of land listed in his will's "Schedule of Property." Over two hundred years later, roads and highways, proximity to both the city of Washington and superb scenery, and, more recently, Dulles Airport still make Loudoun land valuable and drive speculative land purchases there. And the Virginia scion's habit of speculating in land, as we shall see in later chapters, inspired others.

Forty years after the Fairfax decision, in 1999, when the population of Fairfax was approaching a million people, I was living in Boston with my husband and children. My mother began to send me clippings from the *Washington Post* about a zoning battle in Loudoun County that was eerily reminiscent of Fairfax in the 1950s. In fact, according to the *Post*, bumper stickers were cropping up in Loudoun that read, "Don't Fairfax Loudoun County!"

Loudoun County is part of the third suburban ring around Washington, D.C. While Arlington and then Fairfax exploded with growth at midcentury, Loudoun County, with about 20,000 inhabitants, continued to host horse farms on hilly pastures that roll toward the mountains. But Loudoun's growth rate slowly increased in the 1960s, after Dulles Airport was built, and after the 1970s it took off. In the 1990s, like Arlington and Fairfax before it, Loudoun's population nearly doubled, from about 86,000 people to about 170,000—or about tenfold the population of a century before. Like Fairfax in the 1950s, Loudoun had become the second-fastest-growing county in America on its way to achieving the "fastest-growing" status a couple of years later, a position it held from 2000 to 2003, having overtaken Clark County, Nevada, home to Las Vegas. Just as in Fairfax in an earlier day, most of the newcomers to Loudoun were concentrated in the eastern third of the county, closer to Washington, with the western two-thirds of the county remaining rural.

There are differences, too, though. Thanks to trends such as smaller families and larger homes, the growth of Loudoun's population has taken more land than the population booms of the Arlington or Fairfax during similar, earlier periods of growth—using bigger homes to house fewer people. In the 1990s, builders constructed thirty thousand new homes in Loudoun, holding an average of 2.6 people, while Fairfax's 1950 homes held an average of 3.7 people.

CARVING UP HEAVEN

In November 2002, I returned to Loudoun County for the first time in twenty-six years. On the plane, I took out three folders of neatly folded clippings that I had first opened in Boston, along with some notes from conversations with Loudoun officials and printouts from the extremely professional Loudoun County Web site. I reread the stories of rapacious growth and of Loudoun's struggle to direct that growth. As Loudoun's head planner, Julie Pastor, had already told me by phone, county residents had spent the 1990s electing a Board of Supervisors that would pass new zoning measures to help preserve some of Loudoun's historic beauty and character and slow the pace and the burden of growth upon the town. The supervisors' slow progress toward this end was generating headlines.

However, tensions had arisen over the proposed regulations. Some Loudoun residents took up positions so vehement that one resident quoted in the *Washington Post* characterized the situation as a near "Civil War." The county held a nearly interminable series of nearly interminable public hearings. Homeowners, developers, newcomers, and old flocked to air their viewpoints or to draw attention to details of the proposed zoning plan that affected them. From this great cacophony of opinion, the supervisors tried to craft a zoning plan that reconciled the chief interests of residents and developers: Developers were eager to develop and some residents were eager to capitalize on the rising value of their homes. Other residents, an apparently greater number, wanted to preserve some of the rural character of their surroundings, which are, most people agree, beautiful countryside.

When my mother picked me up at Dulles Airport and we headed for Leesburg, Loudoun's county seat, the new access highway on which we sailed along was the first hint of the brave new world before me. I was used to visiting Arlington once or twice a year to see my mother. I never found the same town I left. Especially with the construction of the Washington Metrorail line through Northern Virginia in the 1980s, development had accelerated, concentrated in a few areas around the Metrorail stops. But Arlington had transformed through a steady accretion of small changes. After the townhouses came more office and apartment buildings clustered around Metrorail stops, taller every decade.

Heading west from Dulles Airport, however, I saw the results of a different kind of growth. By purchasing one or two neighboring farms, developers could amass properties for vast new subdivisions of large houses crammed cheek-by-jowl in endless rows. Of the ninety-eight subdivisions completed in Loudoun County in 2001, the largest three featured an average of three thousand homes.[1] The more development gobbled up land, the more expensive the surrounding land and housing became. Loudoun's average single-family home price was already well out of reach of average Americans.

Entering Leesburg, we passed through the town's new corona of one-story shopping centers, stores, and supermarkets set back from the large arterial road on which we traveled. The historic center of the town remained intact, though surrounded by sprawl.

The next day, my mother and I drove west in search of the Loudoun of our memories. We were seeking some of the old scenery and some of the small towns we had driven through many years ago. As we headed west from Leesburg, a new kind of subdivision appeared—basically a pasture with a Macmansion crowning each small hill and sometimes an old farm house among them. These were the larger-lot subdivisions.

When we drove into a stretch of one-story shopping centers and gas stations called "Round Hill," I lost heart. Round Hill had been a beautiful, quiet village on our earlier drives many years ago. But this agglomeration of development that had sprung up on the *new* Route 7 had not displaced Round Hill. When we found the old Route 7, we found the old Round Hill, with its small gathering of stone houses and pastures unscathed. The farther west we drove, the more prominently the Blue Ridge Mountains presided over gently undulating terrain untouched by bulldozer or backhoe.

The human and political sides of the changes we witnessed were on display at the public hearing on the new zoning ordinances that I had come to attend. The dramas were small and personal. They revolved mainly around the self-interest of parties at cross-purposes, though some were very poignant and some of the public officials occasionally exhibited an imperfect, everyday kind of heroism found in those who constantly endure tedium and ingratitude in service of the public good.

Similar dramas play out across the country every day. And because they are a kind of punch line to the long history of American land use, we must look at this particular drama closely enough to understand its importance in that history. This requires of us a little of the patience of de Tocqueville, who understood that local politics, like state and national politics, decide who gets what in America.

The *Washington Post* articles had portrayed a town wracked by conflict over the pace of development and on the verge of implementing radical new zoning laws to counter development pressures. The townsfolk indeed seemed to have strong opinions of all sorts on the new zoning ordinances, and the supervisors, having already approved the policies behind the new ordinances, did indeed seem poised to pass zoning ordinances that would limit new housing construction in the western portion of the county to lots of twenty or fifty acres—less if the houses were clustered to leave larger areas of open space.

However, the ordinances also contained many conventional concessions to developers. For example, the county proposed annexing large unincorporated areas to existing towns, to make those towns responsible for providing infrastructure for services for new development on the annexed properties. Although I had spoken to Leesburg residents who opposed this "developer-driven" feature of the proposed ordinances, as they called it, a number of residents who testified against this feature at the hearing lived on lands proposed for annexation. These speakers did not want to lose the current autonomy of their neighborhoods, and they did not want to be surrounded by the new development that was certain to occur as a result of the annexation or to be forced out of their homes for new development.

Speaker after speaker testified under the brilliantly successful three-minute time limit imposed by the supervisors, and each had a story about how the proposed ordinances would affect them, their families, and their heirs by affecting the value of their home and land. These are the topics land use laws seek to mediate. As at the earlier hearing in Mashpee, the tensions between two kinds of protection—protection for individual enterprise and protection for the interests of the community—were thrown into impressive relief in the Loudoun County hearings.

A number of developers and representatives of national development organizations such as the National Association of Industrial and Office Properties spoke against the constraints imposed by large-lot zoning. They spoke of the need for jobs and taxes that could come with commercial development or the housing to be provided by residential development, depending on their interests.

More speakers were homeowners. One or two homeowners assured the supervisors—seated on a long podium at the front of the room—that their lives and the lives of their aging parents and heirs would be ruined if they could not sell their land to a developer for a large profit. Some speakers, a greater number, assured the supervisors that their lives, the lives of their aging parents, and the lives of their heirs would be ruined if the land that their family had held for generations was spoiled by development. Typical was Lillie Richardson, who spoke before the board a few days later on November 6, as hearings continued. Ms. Richardson appealed to the supervisors to keep her parents' property—surrounded by other residential development—residential rather than zoning it industrial.

> This is my parents' single, by far, largest investment of their entire lifetime. That's why I'm just sort of, I hate to say I'm pleading with you but it's a big deal to them. This is the biggest investment, biggest asset that they have ever owned, and the zoning is the key to the whole thing. It's either not worth a nickel or it's very valuable. So, I would just ask you—also, it seems reasonable. Even if I didn't care about them or their interests, I mean, this is a piece of property, 40 acres worth of which is sitting smack in the middle of a gigantic subdivision which is totally residential. It seems almost, frankly, silly that it shouldn't be residential. So, that's it. Thank you for your time.[2]

All these speakers were asking for protection of different aspects of their property rights or invoking whatever community interest best served their own interests. Some were asking for the traditional zoning protection of the value of their land, and some of them badly needed to protect a home that was the only source of whatever wealth they had. Some people were seeking to maximize their resale profit. Some were asking for another kind of protection—of their right to enjoy the property, the views, and the peace they had purchased or inherited or grown up with. Ironically, Americans have successfully sued the pig farmers on which their suburbs encroach for affecting their air quality or simply the enjoyment of their property, but American homeowners have not been able to sue anyone over the ruined quality of the air they breathe as automobile traffic and congestion increase around them. Still, the recognition of the toll taken by pollution and congestion and other side effects of carelessly managed growth is increasing. We understand as never before exactly how, if I sell my property to a developer, your life as my neighbor, is likely to change.

There is an ever-louder call to redraw this line between individual and community benefit. There is a growing sense of a community's obligation to intelligently manage growth and protect its residents and resources from growth's worst effects, such as pollution, congestion and crowding, and rising property taxes. Loudoun residents of long standing complain particularly about the rising property taxes necessary to support new development. Newcomers complain about not receiving the services promised by developers because the county cannot keep up with the pace of growth in the building of schools and provision of other kinds of services.

In Loudoun, as the supervisors prepared to pass the new large-lot zoning ordinances to protect Loudoun's rural west, there was a sense of progress. Since the 1959 Fairfax decision, some evolution in the thinking of the Virginia judiciary had occurred, giving rise to confidence that Loudoun's new regulations would withstand legal tests. After all, a number of more recent zoning-related legal cases in recent Fairfax and Loudoun history supported this optimism. When in 1985 a homeowner sued the Fairfax supervisors because they would not allow him to divide his 1.5-acre lot for resale, the case again landed in the Virginia Supreme Court. This time the court responded that the issue was "fairly debatable." "Under such circumstances," they declared, "it is not the property owner, or the courts, but the legislative body [the board of supervisors] which has the prerogative to choose the applicable classification."

In an earlier case in 1980, a circuit court approved Fairfax's right to protect its water supply by zoning for five-acre lots around the Occoquon Reservoir. These two cases, and Loudoun's long history as a rural community, gave hope and precedents to Loudoun County officials and staff that their plan for ten- to fifty-acre lots in the western part of the county was politically feasible.[3]

In addition, the U.S. Supreme Court had recently come down on the side of the community in a suit brought against the Tahoe Regional Planning Agency of California and Nevada, objecting to its environmentally based land use restrictions. Though the Court decided the case on a technicality, its opinion clearly supported the land use regulations, as Supreme Court decisions had done throughout the twentieth century.

In January 2003, after years of public testimony on the matter, Loudoun County officially adopted the proposed new zoning ordinances. The retribution of development and property rights advocates was swift. Two hundred lawsuits greeted the new plan. One agitated developer-lawyer-property owner filed fifty of the two hundred suits. Another fifty were dropped as the new plan and ordinances became better understood. Eventually, about one hundred were combined in various categories of shared interests to go forward.

In the words of Michael Laris, of the *Washington Post*, "development interests . . . stood to lose tens of billions of dollars in hoped-for business if county building curbs were allowed to stand. Industry representatives and their supporters filed hundreds of lawsuits and made hundreds of thousands of dollars in campaign contributions to help their allies win local elections."[4] That money helped to run a deceptive campaign in the fall of 2003, while the lawsuits were working their way up to the Virginia Supreme Court, to replace the supervisors who had tightened growth controls. New conservative Republican candidates promised to retain the county's new zoning. But once in office, the new supervisors made their intentions clear.

Even before litigation reached the Virginia Supreme Court, the new conservative Board of Supervisors began to settle suits and overturn growth controls and voter mandates. "What passes for politics in Virginia . . . ," sighed Bob Lazaro, then an aide to supervisor Scott York, the board chairman who had been cast adrift by the aggressive new progrowth supervisors.

First, the new supervisors, led by prodevelopment supervisor Steven Snow, canceled the program the county had instigated a few years before for the purchase of development rights for crucial properties—a common tool for saving land. Then the supervisors settled a suit brought by a developer who had served as Snow's campaign treasurer.

Next, the board considered trading voter-approved parkland to a developer who offered to build a smaller park on another piece of land nearby. The developer behind the deal, Leonard "Hobie" Mitchel, served on both the Commonwealth Transportation Board, charged with siting new roads in Virginia (although he eventually resigned that position under media pressure), and the Loudoun Sanitation Board. Mitchel had earlier proposed an amendment to Loudoun's official plan giving the Sanitation Board the authority to decide on proposals such as his own.[5] Though the board held off on the Mitchel proposal as it garnered bad publicity, two conservative members of the Board of Supervisors had helped to appoint themselves to the Sanitation Board to aid in development decisions. Supervisor Snow became the Board's chairman while still a supervisor.

The activities of the new supervisors raised the stakes of the anticipated Virginia Supreme Court decision, as both sides of the battle awaited vindication. In March 2005, the court handed down its decision: a virtual repeat of the 1959 decision on Fairfax. On an even more picayune technicality than in 1959, the court threw out the large lots. The court found that, throughout Loudoun's laborious citizen outreach efforts and hundreds of hours of public testimony, Loudoun had not adequately specified in public notices the exact parameters of "Western Loudoun," the area for which the larger lots had been designated. The large-lot rural zoning appeared to be doomed.

However, just as the progress of the growth-control movement had galvanized the prodevelopment interests in 2003, the audacious actions of the new board and the decision of the Virginia Supreme Court brought a backlash. Loudoun residents called and e-mailed their supervisors. A Mason-Dixon poll showed the residents' dissatisfaction with the rapid pace of growth. Aide Bob Lazaro tired of the new supervisors' aggressive politics and left to work for a nonprofit conservation group. Miraculously, within a few months of the court decision, a compromise among five of the county's nine supervisors, apparently bowing to public opinion, saved much of the zoning.

THE BIG PICTURE

Loudoun County is only a part of the picture of continuing growth in Northern Virginia and only one front in the American war over property rights. Both sides—community and development advocates—have seen successes and setbacks around the country. In 2005, community interests lost when a voter referendum overturned Portland, Oregon's famous Urban Growth Boundary, which limited sprawl outside the city. But in other instances, visionary programs thrive. A number of communities—Boulder, Colorado; Nantucket and Martha's Vineyard in Massachusetts; and Suffolk County on Long Island—have all instigated programs to purchase land in the last forty years. They dedicate all or portions of sales or real estate taxes to the purchase of open space and in one case to affordable housing.

Further, in the last national election in November 2004, 161 conservation measures made it to the ballot, and three-quarters were approved,[6] even though such measures require either public funding through taxes or bond issues. As mentioned earlier, the U.S. Supreme Court continued to uphold community interests with its 2002 support of strict regulations to preserve Lake Tahoe. Again in 2005, it supported the right of New London, Connecticut, to redevelop a small residential area, although since New London planned to hand the properties over to a private contractor for redevelopment, the decision blurred the line between community and private interests.

Harvey Jacobs, a property rights expert who teaches at the University of Wisconsin, offers the reasonable approach brought by those with historical perspective. He points out in both his lectures and his essays in *Private Property in the 21st Century* that battles to define property rights have been going on in America for centuries and will continue, as the "bundle" of rights that we consider property rights continues to change with society.

Of all these historical changes, three seem chiefly responsible for framing the modern debate differently than the founders did. First, the popularization of speculation throughout American history, traced in this book, has altered the economic dynamics of the modern battle and has motivated many property owners to reach for the highest value that speculative development can confer on their land.

Second, and related to the speculative trend, many observers also consider our concepts of community and civic responsibility diminished and observe that property owners are more likely to ignore the burden created on the community—that is, on other property owners—when they sell their land to a developer. American law does not guarantee American property owners the right to reach for maximum profit without consideration of the contribution the community has made in adding value to their land through schools, transportation improvements, utilities, and other services—as some property rights advocates believe—but this seldom inhibits our tendency to do just that.

Finally, the exponential growth of population has changed the setting of the debate. The more of us there are, the larger the community becomes as it vies for the interests of an ever-larger majority against interests that serve fewer. The value of land as a national resource and public asset increasingly competes with the value of land as a privately-held commodity.

These are changes we must trace through history to understand current attitudes toward the use of land in America. We can blame planners and developers for the sprawling, resource-wasting nature of our settlements, but we must also examine our own predilections as consumers and investors. The desire to own a home and yard is powerful in America and understandably so. A nest of one's own with a grassy space for the children is a dream as seductive as being privately conveyed in an automobile to the destination of your choice whenever you like. And yet, these addictive dreams of American life, multiplied by three hundred million, are using up land and polluting resources at a mind-boggling rate that leaves less to go around for everyone else, including our children.

So we must look at our history and ask how the dream evolved. How was the settler's iconic virtue transferred to the suburbanite sitting on his or her small carpet of lawn? And how was the economic power of a nation invested in the constant grinding of the bulldozer and backhoe?

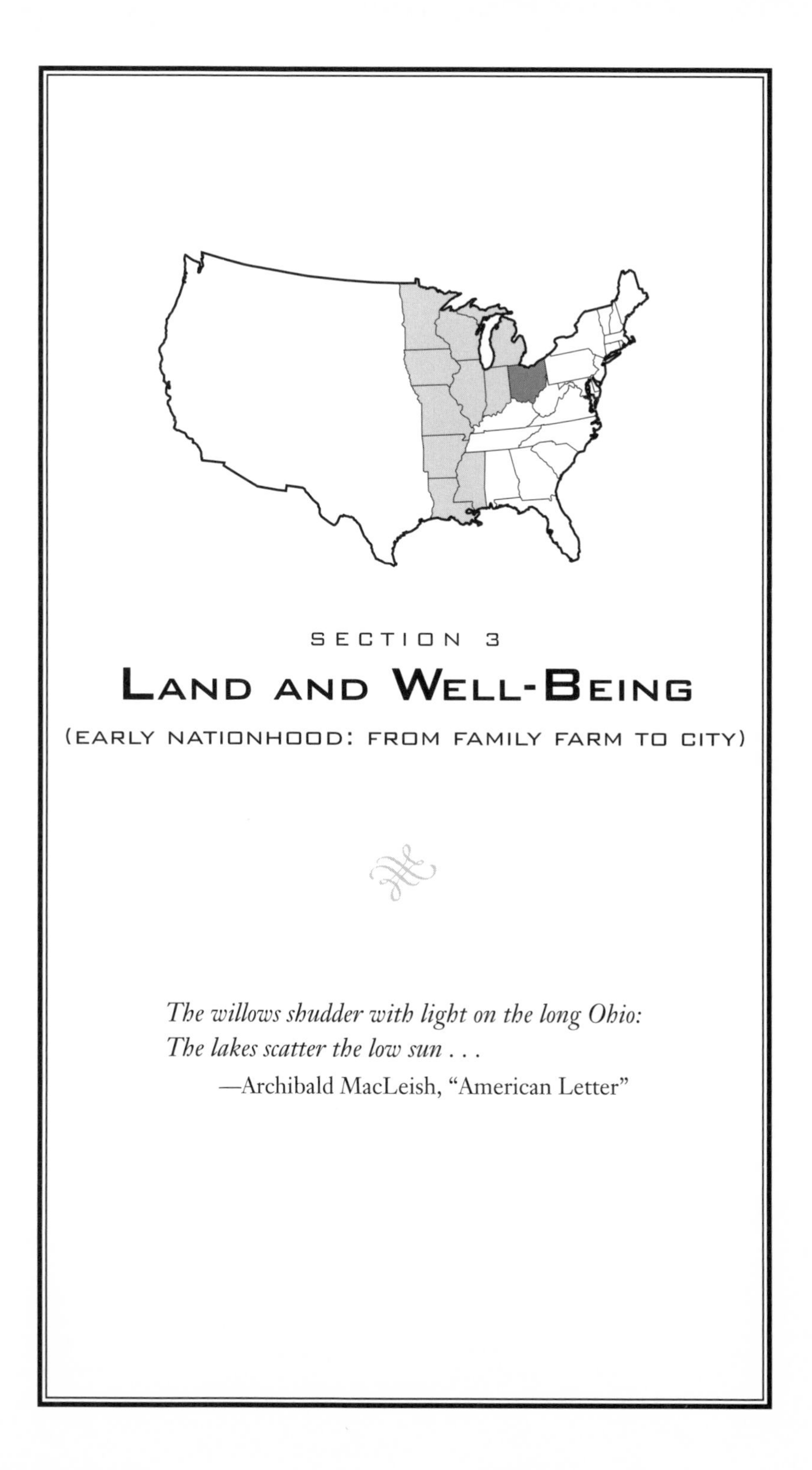

SECTION 3

LAND AND WELL-BEING

(EARLY NATIONHOOD: FROM FAMILY FARM TO CITY)

The willows shudder with light on the long Ohio:
The lakes scatter the low sun . . .

—Archibald MacLeish, “American Letter”

CHAPTER 6

OWNING OHIO

Search for a congenial or at least a tolerable relation between man and land has gone on throughout recorded time, for that relation largely shapes the relationship of man to man. Societies can become homogeneous or polarized, depending mainly on whether land-ownership is distributed widely among the many, or concentrated in the hands of the few.

—Paul Schuster Taylor

The streams of Arlington, Virginia, where I grew up, run through ancient gullies toward the Potomac. Early on, the county preserved many of these wooded ravines as parkland. One such park runs just downhill from my mother's home in a leafy, 1960s subdivision. You can follow this stream—crossing on rocks or old logs when the path gives out on one side—for a half mile as the ravine steepens and, finally, the stream tumbles into the Potomac. From there, a quick scramble up the bluff rewards you with broad views of the river.

Across the river, on the opposite bluff, are the spires of Georgetown University, with the old brick row houses of Georgetown—tiny in the distance—rambling eastward toward the heart of the city. To the left, or west, the scene looks wild. Just beyond Chain Bridge the river courses through a narrow, rocky channel between heavily forested banks. This is the fall line of the Potomac River, the first rapid a ship would encounter coming up the Potomac from the Chesapeake Bay, many miles east. For that reason, it is the first leg of the Chesapeake and Ohio (C&O) Canal, designed to connect Washington and Alexandria to western markets. In reality, the canal got no farther than Cumberland, Maryland, whence came coal to Washington.

The canal lies just out of sight behind the trees on the opposite riverbank. When I perch on this particular bluff, I know it is there only because my father took us often, as children, to walk the level dirt paths along the canal or to canoe its lazy bends. Sometimes we rode the replica canal boats, tugged along gently by mules so slow they seemed to parody the very idea of transportation. But, the park guides assured us, some passengers did choose to travel this way, in the company of many tons of goods.

I did not learn until many years later that the canal was a late incarnation of George Washington's dream of opening the Potomac to westward navigation and achieving a connection to the Ohio River or that the realization of such a scheme would have increased the value of Washington's lands, both in the west and at Mount Vernon. Washington wrote often, especially to Thomas Jefferson, of his concern for the need "to open all the communications, which nature has afforded, between the Atlantic States and the western territory" in order to encourage trade and common economic and political interests among the states and expanding territories to the west.[1] Washington himself obtained a license to improve the Potomac and founded the Potomac Navigation Company, which undertook skirting canals around the falls. Work on the C&O Canal, which parallels the Potomac, actually began in the 1820s, during the national canal era, after Washington's death and the demise of his company.

Downriver, beyond Georgetown, lies more of Washington's legacy: ceremonial Washington with many of its serene marble memorials arrayed along the river. If you walk in that direction, strolling along the paths between the river and the scenic George Washington Parkway that winds beside it, you will come abreast of the Washington Monument on the other side of the river. If you walk several miles farther, past National Airport and Alexandria, you will come to Mount Vernon, the remaining core of George Washington's family estate.

In other words, you would hardly be able to forget for even a moment the name or legend of our first president, lobbyist for the siting of the new federal city here on the Potomac, and commissioner of the L'Enfant plan for the city of Washington. Despite or because of his ubiquitous name and ghostly presence in the area, George Washington remained, in my mind, a marble icon, until well after I left the Washington area. When I eventually learned more about him, through histories and through his letters, it was his interest in the lands of the Ohio Country that lay far up the Potomac and into the wilderness that finally gave him life and breath in my imagination. He, too, had looked upriver at the wilder reaches of the Potomac and wondered what lay beyond. And he had gone to see. In his blend of rare qualities, George Washington has little in common with most of us. But in his love of the land, his curiosity about western lands, and his belief that they could enrich both himself and a nascent nation, George Washington resembles many more ordinary Americans throughout history.

Washington's early westward journeys—the first at about age sixteen—were as a surveyor's apprentice and then a surveyor. He went twice to survey the western lands of Lord Thomas Fairfax, a mentor to the young Washington, who was fatherless after the age of eleven. Lord Fairfax was also a distant relative through Washington's half-brother's marriage to Fairfax's cousin. Fairfax's lodge, near present-day Winchester, Virginia, was already a few days' ride west of Mount Vernon, a ride that took Washington through Loudoun County, where some of the fine stone houses in Leesburg date to the era of Washington's travels. Fairfax's western lands lay in and around the Shenandoah Valley, where Washington picked up more and more surveying work as settlers arrived.

The relatively accessible skills of surveying helped many a clever young man in America to acquire land. As he surveyed, Washington purchased good land when he saw it, amassing twenty-three hundred acres by his twentieth birthday. His appreciation for the economics of speculation is embodied in a much later letter concerning the potential terms of sale or lease of his extensive Ohio acreage. He explains a reluctance to sell: "I well knew they [his Ohio lands] would rise more in value than the purchase money at the present time would accumulate by interest."[2] Land, in other words, was more valuable than money—at least land in an area opening to rapid settlement.

Washington embarked on a greater western adventure in 1753. King George, with the encouragement of the British Board of Trade, had authorized Virginia governor Dinwiddie to grant to the newly formed Ohio Company of Virginia half a million acres of land in the Ohio River Valley, with the stipulation that they plant a settlement of one hundred families somewhere on their tract. The wealthy, prominent Virginians of the Ohio Company were eager to expand the colony and thereby their own access to speculative wealth. Washington's half-brother, Lawrence, and another half-brother, Augustine, had been members. So it is not surprising that, when intelligence reached Virginia of increased French activity in the Ohio Valley, Governor Dinwiddie dispatched twenty-one-year-old Major Washington—who had volunteered for the job—to the French Fort LaBoeuf near Lake Erie. His mission, in the words of historian Francis Jennings, was to "tell the French to go away."

Washington's travels took him along an old Indian path going north from Winchester, through the pass at Cumberland, Maryland, and via the confluence of the Monongahela and Allegheny rivers to form the Ohio. There Fort Pitt would soon be built (at Washington's suggestion) to face the French Fort Duquesne, and there Pittsburgh would one day stand. Passing up the valley, Washington remained east of modern-day Ohio for most of his journey. This land, mostly in western Pennsylvania and West Virginia, was the threshold of the Ohio Country—a coveted land that ignited contentious negotiations between the Virginians and the Pennsylvanians as to who would claim it and also between the French and the British, not to mention the tensions among the Native Americans and colonies over this region.

Jennings and others see the expansionists in Virginia and the British Board of Trade as igniting the French and Indian War of 1754 with this one land grab. However, the demand for land was mounting throughout the English colonies, and pressure to settle the Ohio Valley was widespread. In Massachusetts, Connecticut, and Virginia, no good land remained for the small farmer starting out. Pennsylvania was actively negotiating for land with the Indians of Ohio. Throughout the colonies, settlers, as well as speculators, yearned to cross the Alleghenies.

When the French and Indian War started, Governor Dinwiddie provided two hundred thousand acres of Ohio land to be offered as a shared incentive to any Virginians who would enlist to fight. The later division of this land was the occasion of an unattractive episode in Washington's career as a speculator, as outlined in Jennings's *Empire of Fortune*, which quotes a 1767 letter from Washington to his land agent, William Crawford.[3] Although the British Proclamation of 1763 granted Washington five thousand acres for his services, the distribution of the Virginia military lands promised by Governor Dinwiddie was held up, and Washington grew impatient. In the year of the proclamation, Washington instructed Crawford to "Secure some of the most valuable lands in the King's part, which I think may be accomplished after a while, notwithstanding the proclamation, that restrains it at present."

After a few more years, Washington petitioned Virginia governor Botecourt to reinstate the military land claims—for himself and his men only. Between 1772 and 1774 Washington took a large share of the bounty land—about twenty-three thousand acres or 11 percent—by claiming his allotment of almost eleven thousand acres, in a seventeen-mile chain of riverfront parcels, and purchasing military land warrants from other veterans.[4] By assembling a large patchwork of smaller parcels with the help of a friend in the land office, and by trading and purchasing others' military land scrip, Washington created "an illicitly large tract of land . . . to the detriment of Washington's Virginia comrades in arms for whom these lands had been intended."[5] Washington took the very best lands. One of three plats of his lands along the Ohio River shows over forty consecutive waterfront lots joined together.[6] As part of the five thousand acres granted him in the British Proclamation of 1763, Washington later claimed a similar string of riverfront parcels along the "Big" or "Great" Kanawah and also land near present-day Charleston, West Virginia. Riverfront land not only was flat and fertile—therefore productive and easily farmed—but still held a great advantage for the transportation of goods in a land where roads were few and rough. This pattern by which the best land disappeared first was later repeated by others in the Virginia Military District of Ohio, where much of the hillier land, or land remote from riverways, was never claimed.[7]

Eventually, Washington ended up with over thirty-two thousand acres along the Ohio and Kanawah rivers.[8] (Some came as a reward for his leadership of the Continental Army during the Revolution.) In a letter of 25 February 1792, Washington replied to one Reuben Slaughter, whose land claims overlapped with his, that "in the year 1769 or 1770, there was a special order of the Governor and Council of Virginia for reserving all the lands on the Great Kanhawa [*sic*], to satisfy the military claims of myself and others of the first Virginia regiment;

that in 1770 I was myself on the Great Kanhawa [*sic*] with the surveyor to look out the land for the military claims; and that my patent for the tract you speak of has been in my possession for many years."

As he began to set his affairs in order in the year leading to his death, and to sell some land parcels, Washington's holdings were vast. He had expanded Mount Vernon from the approximately two thousand acres he inherited at the time of the death of his half-brother, Lawrence, to eight thousand. In addition to his lands along the Ohio and the Kanawah rivers, he held land near present-day Cincinnati, plus lands in New York, Pennsylvania, Maryland, Kentucky, and several counties of Virginia, including Loudoun, for a total of about fifty-three thousand acres.[9]

Washington mentioned in letters after the French and Indian War that he felt ill-paid for his service to Virginia, part of which had been volunteer (his trip to Fort LaBoeuf) and the rest of which—military service as an officer in the French and Indian War—had paid a "trifling." Nonetheless, in taking it upon himself to provide his own reimbursement by grabbing the choicest Ohio lands, Washington showed himself to be a Virginia aristocrat who shared with many of his peers a healthy sense of entitlement and a keen appreciation of the value of land.[10] Nonetheless, land acquisition was a sideline for Washington and not a career as it was for some of the members of powerful Virginia families.

In his history of colonial land distribution practices, Daniel Friedenberg has unkind words for the land acquisition practices of the Virginia aristocracy, which had counterparts in other colonies. (The Allen family of New Hampshire, for example, grabbed tens of thousands of acres via various administrative posts, including the governor's office. And land ownership arrangements in parts of New York were nearly feudal.) The first land distribution system to be corrupted in Virginia was the headright system. According to Friedenberg, "The system was subject to flagrant abuse by means of issuing head rights to nonexistent persons, and by 1650 estates of many thousands of acres were appearing."[11] Prominent families, already rich in land, were heavily represented in Virginia's governing council until it disbanded after the Revolution, and they used their influence with the governor and other administrators to award themselves still more land. When the British Board of Trade wanted, in 1696, to find out why Virginia was not growing at the same rate as the northern colonies, Edward Randolph reported that servants were no longer willing to come to Virginia because "the members of the [governor's] Council engrossed all the land for themselves."[12] Barbados had suffered a similar fate earlier, during the seventeenth century, and had sent ten thousand emigrants seeking land elsewhere—many indentured servants whose terms had expired—helping to settle lower North Carolina, South Carolina, and parts of Georgia and Florida.

The disenfranchised of Virginia moved mainly south and west. They first settled northern North Carolina. Later, they surged into Kentucky and the Ohio Valley, taking on the risks of Indian attack, isolation, and the hard work of land clearing in exchange for the best plot they could find. Abraham Lincoln's grandfather was among them. Though his father had given him 210 acres of "the best soil in Virginia," the West beckoned to this Lincoln, also Abraham, possibly through news from distant kinsman Daniel Boone.[13] Whether, like his contemporary George Washington and many others, the young Grandfather Lincoln longed to hold a more impressive estate and to share in the wealth he saw accumulating around him, or whether he felt drawn by the promise of a new society or the lack of society that might be possible in the uncrowded abundance of the frontier, we can only guess. In any case, his initiative was rewarded with over five thousand acres of the best Kentucky land—and an early death in his own cornfield at the hands of Indians who, presumably, had also been fond of his five thousand acres.

The colonial aristocracy had little regard for those emigrating westward. In speaking of the early settlement of Ohio, the Pennsylvania commissioner of the province, Richard Peters, described settlers from Virginia who "go in Companies, and that Country fills with a mighty bad Crew, rejected by Lord Fairfax, the very Scum of the Earth as every body says."[14]

By the time Washington had finished his western adventures and acquired tracts in the western parts of Virginia, some of my mother's ancestors had found their way west from Tidewater Virginia. My grandfather once tried to explain to me why the midsize farm he grew up on in south central Virginia was unusual—six hundred acres along the Bannister River where his great-grandfather had built the current house to replace one that fire destroyed. My grandfather told me that in Virginia it was more common to find either smaller farms or very large ones. I did not really uncover the reasons for this disparity in landownership until I read more Virginia history.

Despite his speculating in land, George Washington—like his colleagues and presidential successors John Adams and Thomas Jefferson—wrote later in his public career of his hope that the public lands of the United States would "afford a capacious asylum for the poor and persecuted of the earth."[15] John Adams, a New Englander and a man of more modest means than Washington or Jefferson, made a similar statement. "The only possibility then of preserving the balance of power on the side of equal liberty and public virtue is to make the acquisition of land easy to every member of society; to make a division of land into small quantities so that the multitude may be possessed of landed estates."[16]

Jefferson is famous for his earlier-cited remark that "the earth is given as a common stock for men to labor and live on." He followed that sentence with the proposal that, "If for the encouragement of industry we allow it [land] to be appropriated, we must take care that other employment be provided to those excluded from the appropriation,"[17] an idea progressive for his time though regressive for our own. Jefferson was not president but rather emissary to France when he wrote those words. Like many of his egalitarian proposals, including a graduated tax that would burden the rich more heavily, it suggested a government that would take more responsibility for social welfare than had any "developed" nation of that day.

Jefferson's secretary of the treasury, Albert Gallatin, also an owner of western lands, declared in 1796 that "If the cause of the happiness of this country was examined into, it would be found to arise as much from the great plenty of land in proportion to the inhabitants, which their citizens enjoyed, as from the wisdom of their political institutions."[18] Several years later, in 1803, Malthus inducted this observation into the western canon of economic principle when he used nearly the identical words in his *Essay on the Principals of Population.* Lending support to the instincts of Washington, Adams, and Jefferson, Malthus stated that the "happiness of the Americans depended much less upon their peculiar degree of civilization, than upon the peculiarity of their situation, as new colonies, upon their having a great plenty of fertile uncultivated land."[19] Looking at the overcrowding in parts of Europe, he warned that famine, pestilence, and war would reduce populations that outgrew the resources that supported them.

Jefferson and Washington and their associates may or may not have seen how their own large holdings of prime land had helped to displace the poor and disenfranchised of Virginia. They saw instead the abundance of land elsewhere—enough for all—and dismissed the hardships of the pioneer and settler who had to endure the hazards of the wilderness and long journeys to find land. More important, they understood the need to address the settlers' interests in order to maintain their loyalty to the new government. Daniel Boone, after all, had flirted with Spanish citizenship in the far reaches of the frontier. Cheap land was not just a social or economic policy but a politically important one. "It is necessary to find a cement," wrote George Washington, continually contemplating the fragility of the new Union, "to bind all parts of this Union by indissoluble bonds." Land provided that indissoluble bond even as it bred a tremendous spirit of independence on the frontier, just as it had done earlier in the colonies. "In America," de Tocqueville would write, "the people are a master who must be indulged to the utmost limits."[20] In return for their government's largesse, settlers legitimized the land claims of America just as colonists had once legitimized the claims of the English crown, and they contributed to the economic energy of the nation.

To this early national period in American history, we owe the generous rights that government continues to extend to homeowners and landowners, with little requirement for consideration of the greater good. The rights of Americans to immigrate to new, growing communities or to make new homes on virgin, cheaply available land—which shape our communities, or lack of community, today—were affirmed in this era when abundant land fueled both rapid growth and national pride and identity.

Until the twentieth century, making land cheaply available—or free after the Homestead Act of 1862—was the only social welfare policy of the federal government. It served brilliantly to provide home, property, and livelihood to many—though the larger share of the population remained disenfranchised.[21] Those who managed to obtain free or cheap land received a direct share in the resources of the continent. In that day, the federal government, which acquired over two million square miles of additional land between 1803 and 1867 through purchase and conquest, could afford to be generous. Today, we have a different economy, a more sophisticated social understanding, no such abundance of land, but enormous national wealth. We now find that the rescue of the poor and lower middle class has become a more complicated and burdensome enterprise. Perhaps our modern failure to fund social welfare more comprehensively or to fund education or health-care programs, except minimally and grudgingly, is to be expected of a society for whom surplus land originally substituted for social policy.

The democratic ideals expressed by these early presidents and some of their colleagues influenced or coincided with federal policy. However, the entrenched practices that allowed the affluent and ruthless enormous advantages in competing for resources—practices exercised by Washington himself—also continued. And these two trends—democratic principle on one hand and speculative land-acquisition practices favoring the wealthy on the other—continued strong in American culture and socioeconomic activity as the pressure for land—as a means to livelihood or wealth—drove people westward. American history is the story of competition between these two tendencies. And to follow them, to follow these two sides of George Washington's nature a little further through history in search of the winner, is to follow a path that begins in colonial America and leads through the Ohio Country and beyond.

FRONTIER OHIO

Despite the 1763 proclamation line, speculators such as George Washington and settlers of more modest means spilled over the Appalachian Mountains, arriving more frequently after the line was withdrawn several years later.

Before the end of the Revolutionary War, westward-bound settlers had a new government to which to appeal their land concerns. In the words of Benjamin Hibbard, "The best lands on the Ohio side were probably being cultivated by people who lived on the Virginia side of the river. In 1778–79 illegal settlements were being made in Ohio and in 1780 those that wished to be law abiding, and wanted Ohio land, petitioned Congress for permission to cross the river."[22]

Settlers were already swarming the fertile lands of Kentucky, whose native people had suffered a momentous defeat at Point Pleasant. Settlement in Ohio was initially slowed by the powerful Ohio confederacy of native tribes. By the first U.S. census in 1790, seventy-three thousand white settlers occupied Kentucky, which soon became a new state. By comparison, at around the same time, only about four thousand are assumed to have been living in the Northwest Territory.

All this activity preceded any land distribution policies of the United States, but it prodded such policies into being, as the rich and the poor pressed the new government for access to new land. Virginia was not the only colony to have run out of good land. Massachusetts, with its poor soils and politically powerful shipping and merchant aristocracy, was no place for farmers, as demonstrated by those who joined Shays's Rebellion. Even before the Revolution, or the sale of the last scrap of land in the Massachusetts government's holding, occasional symptoms such as low birth rates (around 1700), emigration, a lower standard of living, and greater enterprise in crafts and small-scale manufacture showed the diminished prospects for the Massachusetts farmer.[23] And though the Mid-Atlantic states, with their highly diverse economies, better supported lower-income residents, they still sent emigrants west.

Substantial immigration from abroad continued as well. Emigrants from the British Isles and elsewhere overseas continued to arrive. By 1796, the road to Kentucky through the Cumberland Gap (not to be confused with Cumberland, Maryland) was wide enough for a Conestoga wagon, which a family arriving by boat at Alexandria, Virginia, could purchase and fill for the long westward journey.[24] Transportation improvements undertaken to improve commerce facilitated immigration as well.

The Eastern Seaboard continued to grow, but it was no place for those longing to own land. Shortly after the Revolution, in 1785, prices around Boston, New York, Philadelphia, Baltimore, and Charleston ranged from $30 to $50 for an acre of "improved land near towns," while the national average was $2.50 an acre. In 1805, a few years after Ohio attained statehood, land prices around those cities ranged from $250 to $300 per improved acre, with the national average at $6.50. Meanwhile, in those same cities, wages had roughly doubled during the same period and then dropped slightly. The cost of most foods had roughly doubled, keeping pace with wages.[25] This phenomenon of intensifying land markets continues throughout the United States today, most conspicuously in cities, raising other costs of living with it.

FEDERAL LAND POLICY

After the Revolution, the new federal government struggled to keep up with settlement. In the interest of the new nation, Massachusetts, Connecticut, and New York ceded their claims to western lands to the new federal government. (Virginia had ceded Kentucky but maintained her claim for military bounty lands in Ohio, as did Massachusetts and Connecticut.)

Even with the war on, Congress began deploying its great arsenal of land. "In August 1776, Congress began by offering land to deserters from the British Army with a special bonus to officers who should induce soldiers to desert with them." The next month, Congress passed legislation granting land to American officers and soldiers according to rank.[26]

In 1780, still at war and still unsure of the outcome, Congress took the first step toward a national land policy. It passed an act that "provided for the disposal for the common benefit of the United States of the territories ceded to the United States; for the formation of states out of these territories; and for the regulation by Congress of the granting and selling of these lands."[27] In 1784, with the land now truly American, Congress put one of its greatest minds, Thomas Jefferson, to work drafting land policy and passed ordinances in 1784 and 1785 detailing the plan for the sale and subsequent settlement of public lands. In 1787, Congress passed the Northwest Ordinance declaring the Northwest Territory—present-day Ohio, Indiana, Illinois, Michigan, Wisconsin, and eastern Minnesota—open to settlement and providing for its governance. The Northwest Territory became a laboratory for the first land distribution policies of the federal government, as it responded to the pressures of settler and speculator.[28]

Many competing land interests carved up Ohio. Pennsylvania bought up the land at its northwestern corner to procure access to Lake Erie. Virginia and Connecticut retained large tracts of Ohio, mainly to fulfill obligations to their veterans. A further large area of the Ohio Country was set aside to provide for the claims of veterans of the war who had received scrip for land as payment for their service (although speculators also bought up large pieces of land by buying up military scrip). A few companies of speculators sought to buy, and succeeded in doing so, large tracts of a million or more acres from the federal government on very advantageous terms. The most successful of these was the Ohio Company of Associates, of Boston, headed by Revolutionary War general Rufus Putnam, another former surveyor, a Boston businessman, and later a delegate to the Constitutional Convention,[29] Reverend Manasseh Cutler. The company ended up with just under a million acres purchased at about eight cents an acre using bought-up military scrip.

A few features of the company show the continuing confusion of land hunger, land wealth, and political power in the new nation: First, the petition for land received the approval, pending congressional action, of Putnam's former commanding officer and now president of the new nation, George Washington. Second, Ohio Company leader Manasseh Cutler had helped to draft the Northwest Ordinance. Third, Putnam's colleagues were officers of the Revolution who joined Putnam in buying up land warrants from other veterans and petitioning the government to accept worthless continental currency in payment for the Ohio land they sought. They ended up with a huge swatch of land on the Ohio River—now Washington County. On their vast tract, they founded the first Anglo-American town in Ohio: Marietta. A look at a map of Ohio from 1803, when it attained statehood, suggests that half the state or more had been granted away to speculative land companies and the settlement of states' claims. This may help to explain the lower rates of land ownership in Ohio compared to late colonial America, to which we will come shortly.

Every land distribution law that Congress passed, no matter its intentions, seemed to somehow leave the settler vulnerable to the speculator. The Land Ordinance of 1784, to which Jefferson contributed, sold land in large town-size chunks, following the New England system of town planting. Washington wrote in support of it.

> Compact and progressive settling will give strength to the Union, admit law and good government, and federal aids at an early period. Sparse settlements in the several new states, or a

> large territory for one, will have direct contrary effects; and whilst it opens a large field to land jobbers and speculators, who are prowling about like wolves in many shapes, will injure the real occupiers and useful citizens and consequently the public interest.

Congress changed the law in 1785, seeing that only speculators had the capital to purchase large tracts. The attempt to refine laws to favor the settler continued throughout the century fighting powerful financial interests. As late as 1910, Teddy Roosevelt would urge Congress to further reform land laws to protect land for settlers against the voraciousness of "great corporations."[30]

But while land laws were generally liberalized throughout the nineteenth century, Ohio was being settled at the turn of the eighteenth century. It joined the Union as a state in 1803, during the earliest stages of Congress's experiments with land laws. Congress established the national land office with branches at Pittsburgh and Cincinnati, both on the Ohio River, and inaugurated the national survey, begun in Ohio in 1785 on the lands thus far ceded by Indian nations, to enable a more orderly sale of land to settlers. Pioneers, however, were perpetually ahead of the survey.

Meanwhile, the federal government continued to try to make terms for the purchase of public lands easier for settlers. In 1800, Congress dropped the minimum acreage to 320 acres at $2 per acre and then to 160 acres in 1804. In 1820, public land prices dropped to $1.25 an acre with a minimum purchase of 80 acres. The requirement for cash up front undercut the generosity of this offering to the settler and, instead, benefited the speculator again. In fairness to the government, the credit system had not worked well, leading many settlers to default and others in debt to new banks that were springing up. Between 1820 and 1832, land laws were adjusted twelve times, but still the speculator benefited. Finally, with the Homestead Act of 1862, the government gave land away for free to those willing to provide proof, over a few-year period, of their intention to settle and improve it rather than sell it. Policies concerning whether squatters had rights to their land and whether credit would be extended to settlers and under what terms all changed continuously through the early decades of nationhood, while Ohio was being settled.

To further bind the interests of Ohio settlers to those of the federal government, President Washington turned to removing the native population of the area. After losing nine hundred soldiers in a bitter defeat there, Washington invested 80 percent of the federal budget in an offensive against the Native American tribes of the region. And, finally, after the defeat of the Ohio confederacy of Indians at the Battle of Fallen Timbers in 1794, and after all the frontier bloodshed that preceded it, major land concessions by the Native Americans allowed settlement to accelerate.

By 1803, when Ohio attained statehood, its population was expanding exponentially—to over two hundred thousand by 1810. Its phenomenal growth slowed only slightly in the next decade as speculators, pioneers, and settlers rushed in, clearing forests and building log homes and barns. By 1820, local leaders were lobbying state officials for help with canals and roads, while state officials, in turn, lobbied the federal government for investment for the construction of canals and other improvements that would boost Ohio's growing economy by allowing farmers to find markets for what they produced.

Canals provided employment to immigrants, such as the Irish who helped build them. They helped to carry new settlers to land. And they provided the convenient connection to markets that helped local farmers thrive. The many streams and riverways of Ohio attracted small industries, too—processors of salt, lumber, leather, wool, and glass—and these were connected to markets. The combined industry of these early settlers and builders of Ohio—whom we shall meet in the next chapter—along with the ceaseless energies of speculators had the inevitable result of making land prices rise along with levels of prosperity. Ohio was no longer the frontier.

By 1838, federal land offices had been established in Illinois, Missouri, Indiana, Alabama, Mississippi, Louisiana, Michigan, Arkansas, Wisconsin, and Florida. Settlers and speculators moved west, following rumors of better and cheaper land and good river access. George Washington and the other founders were dead. Their voices, calling for a democratic division of resources, echoed for a few more decades and then faded. But the speculative investment habits of Washington and his peers lived on and on.

CHAPTER 7

EARLY OHIO AND THE FATE OF THE FAMILY FARM

Who can desire more content, that hath small meanes; or but only his merit to advance his fortune, then to tread, and plant that ground hee hath purchased by the hazard of his life? If he have but the taste of virtue, and magnanimitie, what to such a minde can bee more pleasant, than planting and building a foundation for his Posteritie, gotte from the rude earth, by Gods blessing and his owne industrie, without prejudice to any?

—John Smith

In the years after the opening of the Northwest Territory, hundreds of thousands of people pressed westward. They brought with them a range of abilities conducive to a range of fates. The stories of three men will help us to understand the variations in luck, talent, drive, and starting advantages such as property or capital that sorted people into different economic ranks as the years wore on. These men moved to the frontier, in or near Ohio, in the early nineteenth century. Their stories also show how, as settlement increased, the frontier operated differently to assign rewards and to exact the investment of sweat and sometimes blood that procured those rewards. Duncan McArthur, a poor boy born in New York, became a frontier scout in and around Ohio, a surveyor with a keen eye for good land and a shrewd bargainer. Eventually, he became governor of Ohio and one of the state's largest landowners. Jonathan Hale, a Connecticut farmer, brought his financially strapped family to northern Ohio in 1810, having traded in a much smaller piece of well-situated, inherited land in Connecticut for five hundred acres in Ohio, where the family prospered. Lastly, Thomas Lincoln, father of the future president, lost most hope of good fortune in Kentucky at the age of eight, when his father died and the rule of primogeniture handed his fifteen-year-old brother the entire family estate of five thousand acres.

Although Thomas Lincoln marched into Kentucky with his family as a young child in the 1780s, Duncan McArthur was the first of these three men to make a westward move as an adult. He had been working since the age of seven, when his widowed father hired him out to work on other farms. His physical strength and courage made him a good candidate for frontier life. He trained as a soldier. Around the Ohio Country and Kentucky, he alternated among various positions that involved scouting for Indians or fighting them near the frontier and learning surveying in Ohio's Scotio Valley. He helped to lay out the town of Chillicothe, which was Ohio's first capital when it gained statehood and McArthur's first Ohio home after he and his wife moved from their log cabin with a growing number of children. McArthur was still living in the cabin when he was elected to his first government post, representative in the Ohio General Assembly, in the fall of 1804, but not for long.

Like that great scion of Virginia, Robert Carter, land agent to Lord Fairfax, McArthur also had eleven children. Carter once gave this as the reason why he set about, through his position and connections, amassing his small empire of three hundred thousand acres. By some accounts, McArthur began speculating in land warrants after his children began arriving and perhaps intended to spare them the hardships he had known.

As a surveyor, McArthur had developed a keen eye for choice land, and as a shrewd bargainer, he bought veterans' scrip cheaply both to sell and to use in the purchase of land. In the War of 1812, McArthur served as a brigadier general, increasing his prominence along with his landholdings. In 1823, his district elected him to Congress. In 1830, he narrowly obtained the governorship.

McArthur's story is an archetypal frontier story in more than one respect. McArthur's rise from poverty to prominence demonstrates the equalizing effect of the frontier and the social mobility it promoted. His route to power shows the historic connection between land or resource ownership and political power. McArthur's story concludes with the end of the Ohio frontier and the muting of its equalizing effects. By 1810, with over thirty-five thousand acres, McArthur was one of the four largest landowners in Ohio, three of whom had been surveyors like George Washington before them. The top 10 percent of Ohio landowners in that year owned one-third of all properties registered in the state.[1] Though there was no Lord Fairfax holding five million acres, the early Ohio land companies had held similarly vast tracts and now continued to hold advantages. In Washington County, Ohio, Washington's fellow officers in the Ohio Company of Associates had settled on the banks of the Ohio River, on larger landholdings than their fellow Ohioans.

Lee Soltow, a modern economic historian, chose to study the early Ohio society of 1810 and to ask how American equality was faring as the frontier moved west. It is a convention of economic history that the frontier breeds equality. Economic historians John McCusker and Russell Menard, in *The Economy of British America*, describe the general principles of economic development that govern the frontier: "At first new communities were characterized by low levels of wealth and by the crude egalitarianism of frontier life. . . . Inequality . . . progressed as a few men, often the earliest arrivals [such as McArthur] or those with more capital and good business connections [such as Hale], pulled ahead of their neighbors."[2]

Soltow surveyed Ohio landownership and tax records for 1810, when roughly two hundred thousand people lived in Ohio—about 99 percent of them rural dwelling, as opposed to 94 percent for the nation, with the vast majority working in agriculture. Soltow reckoned that, in 1810, only 45 to 50 percent of the free adult males in Ohio owned land,[3] meaning that a slightly lower number were tenant farmers or farmhands—men starting from nothing rather than with the advantages of early arrivals or those with property, capital, or connections.

If we compare Soltow's figures concerning landownership in the early national period of American history to figures from the late colonial period thirty-six years earlier, we cannot avoid the conclusion that even frontier land was already becoming more expensive and less available than in the colonial era. In 1774, 70 percent of "wealthholders" in the thirteen states owned land. (Wealthholders were mainly white males in both colonial America and early Ohio.)

The lower rate of landownership in Ohio testifies to the increasing clamor of Americans and immigrants for land at the opening of the Northwest Territory and the rapidity and competitiveness with which the best Ohio land was sought and taken up; a society formed; markets and producers connected; and the frontier pushed westward. In the colonial period, when settlement hazards were even greater and markets less numerous and developed, land was often offered as a lure to settlers, either as a headright or, in New England, as a town assignment of a free lot for each founding family or individual.

Free land was vital to the high rates of ownership in colonial America, for stocking a farm was costly. Even when the price of government land was as low as one hundred dollars for eighty acres of the Northwest Territory in 1820—including Ohio—setting up a farm required capital to purchase equipment, stock, and seed. To start a farm, a man would have to save up not only the one hundred dollars for federal land, if he could get it, but another five hundred dollars to outfit the farm and begin to plant.[4] This sounds manageable until we consider wages in that day: around seventy cents a day for laborers on the Erie Canal or nine and a half dollars a month for farmworkers.[5] And land prices were higher during Ohio's early settlement, with land companies selling part of the land.

A HEAD START HELPS

Early Ohio resident Jonathan Hale came from old Connecticut stock in Glastonbury. Jefferson's embargo and continuing tensions with the British may have unsettled his life and finances. And his homestead on the Connecticut River represented the dwindling shares of land divided among siblings over five generations. For Connecticut residents, at any rate, "Migration was the old and established solution for overpopulation, poor soil." In the early nineteenth century, northern Ohio was a popular choice.[6]

With his roughly one hundred acres of inherited land on the Connecticut River, Hale had property to trade on. He also had debts and little cash. He may have sold some belongings, his inherited fishing rights on a nearby island, and possibly other property to pay off his old debts and to finance his family's migration.[7] Because Connecticut had long since sold its Western Reserve in Ohio to a land company, save half a million acres of "firelands" for those whose homes the British had burned, Jonathan Hale traded his land to a member of that company. He moved to Ohio in 1810, the year on which Soltow focused. Hale traded his one-hundred-acre Connecticut homestead for five hundred acres in northern Ohio, mostly uncleared. Starting anew, like other pioneers, the Hales led lives of hard work and privation, eventually crowding into their one-room cabin with five children.

Though their early years were hard, the Hales owned more land than 90 percent of Ohio landowners—a tract three or four times the median size. And they were much better off than non-landowning families. Aside from early hardships and a couple of uneasy years with the Indians, Jonathan Hale found prosperity amid the clover. He discovered he lived on a geological bonanza. A small bed of limestone underlay his land. Limestone nurtures the vast cornfields of west central Ohio. Clover likes limestone, too. And, in a growing community, limestone had a special value. Once Hale had discovered the limestone, and perhaps once his sons were big enough to help with the farming, he burned the stone for lime to sell as mortar while building increased in the towns around him, continually increasing the demand for bricks and mortar. He also made bricks for his own use and slowly built his family a stately three-story home.

After the Ohio Canal was cut near his land, Hale's market and earnings increased along with those of his neighbors, who now had a direct link to markets in and near Cleveland. Prosperous homes gradually replaced log cabins in the area. With Mrs. Hale adding tailoring work to her chores, again as the children grew and helped, the Hales made a good living from agriculture supplemented by cottage industries in the winter months—a combination that had made colonial Pennsylvania the "poor man's Paradise."

Hale's investment—trading his Connecticut land for a larger portion on the frontier and drawing prosperity from it through continual hard work—ensured that his children would never have to work as hard as he had. After his death, one of his sons opened the brick family home to vacationers from growing Cleveland, and another took advantage of the lime-loving clover on their land to keep bees and make honey.

Hale's success, however, was rooted in the even earlier success of his own ancestors homesteading in Connecticut. This economic foundation stood him in good stead—along with the frontier necessities of a strong constitution and good luck. He had come from an old Connecticut family with early claims to good land and an assortment of other assets.

Some of Hale's neighbors were poorer and less fortunate. Occasionally, Hale hired a couple of them—including a luckless brother-in-law to whom he had donated a little land—to help clear or harvest portions of his acreage. But, for most farm labor, Hale was blessed with sons and daughters who grew strong as they grew older.

With the top 10 percent of the 1810 landowners, to which Jonathan Hale belonged, owning one-third of all Ohio properties,[8] a bit less land remained for Washington's "asylum for the poor." The problem of access to land only increased as time went on. While Lee Soltow could celebrate the size of the average plot held by an Ohio landowner in 1810 or 1820 as still many times larger than the average owned in Britain (from which so many land-starved colonists had originally come), these figures would not hold. Every decade brought more people and an intensification of development that, in turn, brought exponentially higher land values and smaller land parcels, despite occasional, brief regional and national panics, recessions, or depressions that temporarily lowered land prices as they do today.

THE CLOSING DOOR

So what of the less fortunate landowners, as well as the nonlandowners of this period—the vast majority of folk? At this point, we turn to the story of Abraham Lincoln's ancestors to understand the much harsher fates the frontier offered and the life of toil that was traded for equity in a piece of land. First, as told by David Herbert Donald, there is Grandfather Lincoln, turning down his father's offer of 210 acres of good Virginia land in order to seek a greater estate, more comparable to his father's, in Kentucky. After Grandfather Lincoln's untimely death at the hands of Indians, his entire estate passed by the law of primogeniture prevailing in Virginia and Kentucky at that time to his eldest son, fifteen-year-old Mordecai, leaving nothing for Thomas Lincoln, Abraham's father. "It required an immense effort for

Thomas," Donald writes, "who earned three shillings a day for manual labor or made a little more when he did carpentry or cabinetmaking, to accumulate enough money to buy his first farm, a 238-acre tract on Mill Creek, in Hardin County, Kentucky." The frontier's equalizing effect foundered when luck or health gave way; we can see how differently the Hales might have lived had they started from nothing. And in 1805, the year Thomas Lincoln married, Duncan McArthur was moving his family into a fine mansion on a hill in Chillicothe.

In 1814, living on his Kentucky farm with his family in a one-room log cabin, Thomas Lincoln ranked fifteenth out of ninety-eight landowners in his county in the size of his estate.[9] Living in conditions that Abraham Lincoln would later call poverty,[10] Thomas ranked, like Hale, above the majority of his fellow residents in property ownership. At this point, we must admit that things did not look good for the majority of pioneer families. America was still a nation of people for whom subsistence farming was a common way of life.

In Kentucky, Thomas Lincoln, faced with a legal battle over the title to his third Kentucky farm, moved his family to the fertile soil of southern Indiana in 1816 and to the task of building a new cabin and clearing forested land. They had heard of good soil there, and it seems safe to assume that by 1816, although public land continued to sell in Ohio, the better lands in that state were beyond their grasp financially. A couple of his wife's relatives came with them, also displaced by the same trend in which wealthier landowners moving into Kentucky, who could afford the legal fees for a battle over land titles, held advantages over smaller landowners, who could not defend their titles. These same large landowners were bringing slaves to Kentucky and establishing plantations with economies of labor and scale with which the small farms could not compete. In Indiana, Thomas would continue a life of grueling physical labor. He would lose his wife, Abraham's mother, to a frontier illness. As Thomas Lincoln and his peers toiled in the midst of such plenty, it is no wonder the American imagination turned toward the concept of speculation and easy money.

PROSPERITY AND DEATH

The frontier moved on. The 1838 report of the General Land Office showed the big sellers to be Indiana and Illinois, on the Lincolns' migratory path, with close to a million and a half acres sold in each. Michigan and Missouri were also ahead of Ohio, in which, nonetheless, 470,000 acres sold. The nation was by then 18 percent urban dwelling and rapidly industrializing, with Ohio in the forefront of industrial progress—and as a result, also near the forefront of American cities in receiving immigrants from abroad to work in its factories. Its superior transportation connections and a wealth of newly discovered industrial minerals being mined in or near the state brought both industry and immigrant labor to Ohio. The fledgling industrial economy, the larger number of people living in urbanized areas, the intensifying competition for land, and the sacrifices of the frontier folk grasping at opportunity, along with the Civil War in the 1860s, all influenced the general well-being of Americans during this period. For the population as a whole, that well-being diminished.

For reasons that are not fully understood, Americans from roughly 1830 to the end of the century were shorter than Americans before or after them. Americans in the 1840s, 1850s, and 1860s died three or four years younger than their American ancestors or progeny.[11] These facts suggest poor diet and other forms of adversity, despite the country's rising agricultural output.

Certainly, inequality rose during this time. Between the Revolution and the Civil War, wealthy Americans—the top ten percent—garnered a much larger share of the wealth held by the American population,[12] causing the per capita income for the nineteenth century to rise without accurately reflecting the fate of the average American. Meanwhile the poor ate from pots with their hands, sitting on the floor,[13] and filled out the ranks of the military, from which height records for the period are taken. Infectious diseases such as cholera stole breadwinners and caretakers from families and shortened lives. Immigration and industrialization contributed to these trends. For awhile, the highest mortality rate in the nation was in the Irish slums of Boston. One economic historian, Lorena S. Walsh, has suggested that increasing urbanization during this period gradually prevented traditional access to the land with which people supplemented their diet by fishing, hunting, or collecting nuts, berries, or mushrooms.[14] And, as the life of Thomas Lincoln shows, the disparity between wages and land prices—the sacrifices required to build "sweat equity" in a piece of land—compromised the health and prosperity of many Americans.

Wages for the common man rose *very* slowly over the course of the nineteenth century. The wages of farm laborers—$9.45 a month plus board in 1818—had not even doubled by the end of the century. On the Erie Canal, in 1828, the scanty wages for laborers and skilled workers were approximately what common laborers had averaged around the country in 1815 ($0.70 to $1.50 a day), and they hovered near the same through 1850. And while the 1828 wage had roughly doubled by 1870,[15] the price of land rose much faster.

Land values skyrocketed after the Revolution. Within twenty years, land values around major cities quintupled or better. In Ohio, land values surged steadily, despite occasional dips when the price of grain or other farm products fell, creating the classic conditions for rising poverty and rising inequality as wages failed to keep pace.[16] As land prices rose, the many costs of living dependent on land prices and rents also rose.[17] In Ohio, where federal land had cost $1.25 per acre in 1820, land in 1880 averaged $46 per acre,[18] bringing the price of an eighty-acre farm from $100 to $3,680. In more populous counties such as Cuyahoga, where Cleveland was burgeoning, average land prices reached $81 per acre in 1870.

THE FATE OF THE FAMILY FARM

By 1840 Ohio, with one and a half million people, was the third most populous state in the Union after New York and Pennsylvania. Its workforce was still 75 percent agricultural but had diversified somewhat, with the next largest group—about 20 percent—in manufacturing. Large, complex enterprises dotted Ohio: machine manufacturers, carriage makers, and steamboat builders augmented the industry of the grain mills, woolen mills, cabinet shops, glass factories, and tanneries of the early nineteenth century.

The rise of land prices followed settlement everywhere. With every new frontier, speculating, fur trading, timbering, and purchasing farmland became easy for a time and then more difficult. And while shrinking land supplies raised prices and made the purchase of a small farm more difficult, other pressures undermined the small-scale farm. By the end of the nineteenth century, a gradual shift of the American economy away from agriculture and toward manufacturing had accelerated, and large commercial farms arose. The arrival of chemical fertilizers, mechanization, improved transportation, and modern agricultural science all contributed to the economies of scale enjoyed by larger and larger commercial farms.[19]

Investments in agricultural machinery could only be justified by large-scale production. By 1847, Cyrus McCormick had opened a factory in Chicago to cater to the large farms of the Midwest and West. "An early McCormick reaper could cut twelve to fifteen acres of wheat in a ten-hour day, while an individual could cut only one-half to three-quarters of an acre with a sickle."[20] The vast, flat stretches of Midwestern and Western land suited machinery and doomed smaller grain producers in the East.

So ironically, by 1862, when Congress, free of the influence of large southern landholders for the duration of the Civil War, finally passed the Homestead Act, granting 160 acres of free land to anyone who would work the land, the smaller farm was becoming impractical.[21] The life of the Hale family became less familiar to Americans. Even the farmer followed trends toward speculation and the purchase of household goods, forgetting his self-sufficient economies[22] and finding himself tempted into debt.

Some of Ohio's prosperity went West. Chicago had already drained some of Cincinatti's wealth by superceding her as the hog market for the Midwest. After the Civil War, grain and cotton prices plunged as large farms farther west came on line. Later, dairy farmers would suffer the same fate. With a Congress hostile to the South or insensitive to the concept of monopoly, unfavorable legislation gave industry advantages over agriculture. Railroads and large distributors could dictate transportation and distribution prices to small farmers, creating another advantage for the large farm.[23] In 1889, the *Southern Cultivator* of Georgia showed its readers an engraving of a long phalanx of horse-drawn riding plows crossing one of "the great farms of Dakota," where a thousand acres were sometimes planted in one crop, such as wheat or flax.[24] On one "Bonanza Farm," as they were called, sixty-one thousand acres were planted in wheat. These enterprises drew foreign investors, just as Americans now operate farms in South America or Australia, where land and labor are cheaper.

During all this time, the gap between wages and land prices continued to widen. Farm tenure, which held steady through 1850, began to decline. Tenant farming, share cropping, and escape to manufacturing jobs increased through at least the mid-twentieth century as farm ownership dropped.

Ironically, Ohio, with its proximity to Great Lakes iron ores, led the manufacture of farm machinery. And, when the new machines stole jobs from agricultural laborers, the factory cities of Ohio received the underemployed rural workers who came seeking jobs.

By 1870, large speculators' tracts had disappeared from Ohio. Log buildings were rare in Ohio by then, but so were wood lots. Transformative economic forces were at work in the relationship between people and their land. The industrial cities, such as Cleveland, were growing. The average Ohio farm had shrunk to half its 1820 size. Farms in urban counties were even smaller—closer to the size of landholdings in the historic Britain from which American colonists had come seeking land. While grain farmers and dairy farmers lost out to large operations, truck farmers close to cities, however, were able to supply fresh produce to the nearby population centers a while longer. But land close to cities was now too expensive to purchase in acreage large enough for farming.

"LAND . . . IS THE SOURCE OF ALL EMPLOYMENT"

The federal government noted the passing of the farm. In 1919, Benton MacKaye, working in the Department of Agriculture, penned a report for the secretary of labor called "Employment and Natural Resources." His pamphlet shows that at least some in the federal government still had faith in the capacity of America's natural resources to provide for the common man. In his report, MacKaye evaluated the American's prospect for economic well-being based on the nation's considerable natural resources and the basis those natural resources offered for employment opportunities. "'Land,' in its broad sense," he declared, "is the ultimate source of all employment. Land in this sense includes much more than agricultural soil. The latter, to be sure, is one of the major resources. But it is only one. In addition to the soils we have the forests, the ores, and the waters."

MacKaye's report addressed the needs of Americans whose lives and employment had been disrupted by World War I, their need for jobs and meaningful careers. But it also discussed the plight of the American worker in more general terms. Development, MacKaye pointed out, was increasing the cost of living. "Wages, to be sure, have been rising as well as prices, but not nearly in proportion thereto. In the 12 years since 1907," he reports, Bureau of Labor Statistics figures "show that real wages, based on cost of food, have fallen nearly one-third. . . . That is to say, where a man earned a dollars' worth of food in 1907, he can today earn only 69 cents' worth."

MacKaye blamed "a decline in the per capita production of food" (even though production per acre was rising) and an urbanizing population now split almost evenly between city and country. He concluded that "the shift of labor from agricultural to urban industries, and the associated falling off in food production, would seem, therefore, to constitute important reasons for the prevailing high cost of living." MacKaye embraced the family farm. He proposed to transfer unemployed or underemployed workers back to the land and to improve the highway and road network that would help farmers transport their goods to market (promoting the trend toward automobile travel that would supercede the railroad system).

Alas, the homestead resurgence was not to be.[25] MacKaye failed to understand all the forces working against small-scale agriculture. And although even a few acres of land to allow a kitchen garden and chickens, such as my mother's family kept throughout her childhood and adolescence, would have improved the standard of living of many Americans, the urban transformation of America into large metropolitan job centers where land for housing was at a premium worked against even this concept.

In 1941, agricultural labor economist Paul Schuster Taylor, who had documented the effects of the Depression in rural areas with photographer Dorothea Lange, his future wife, wrote an article called "Good-By to the Homestead Farm." It mainly lamented the effects of mechanization and environmental degradation on prospects for farm ownership and on labor and noted the rise of the impoverished migrant workers as machines tended more and more—but not all—of the crop cycle.[26] There were still fruits and vegetables that machines could not pick, however, and migrant laborers picked those.

In 1948, the Department of Agriculture reported a sharp decline in farm ownership between 1880 and 1940 and a corollary increase in the number of tenant farmers and laborers. The USDA attributed these trends to mechanization and commercialization of farming, which made larger farms advantageous, putting smaller farms out of business and absorbing their labor, and to reduced land supply. "With the filling up of the frontier," the Department concluded, "the rate of increase in the development of new farms greatly slackened, and replacements of [farm] operators who retired or died became the principal source of accessions to farm-operator status."[27]

As they shrank and vanished, the farms around Cleveland, neighbors of the Hale farm, underwent a transformation similar to those around other large or growing cities. They sold out or converted to truck farms and dairy farms to supply the city rather than competing with the large grain operations out west. Eventually, large dairy farms and corporate distributors put small and midsize dairy farms out of business. Finally, only specialty crops or services that could be sold directly to city dwellers were profitable and only for a few.

These changes were only part of the picture of the changing relationship between the Ohioan, the American, and his or her land in the late nineteenth century. Invisible from rural America, but exerting a greater and greater influence on American life, were the burgeoning nineteenth-century cities such as Cleveland.

Standing in the sun at Hale Farm, inhaling the sweet smell of clover, I contemplated the story that awaited me on the drive back to Cleveland. I would have preferred to stay. The creek, the neglected shoots of corn amid the clover, and the silent home all looked as if the Hales might have left only momentarily, bumping along in their wagon to town. The whole place brought back my Virginia grandfather's stories of the blissful liberties of rural life—picking fruit from the tree, melons from the garden; dropping to the grass to drink from the brook; or skinny-dipping in the river—and the pride he had felt when neighboring farmers asked his father for advice.

I tried to remind myself of the smell of manure that had once permeated the farm, of the bitter cold and the summer heat, of the mosquitoes that had spread fevers among the settlers, and of my own grandmother, an aged farm girl, with her gnarled hands immersed in near-scalding, soapy dishwater on the hottest day of the summer. And yet, I could not help remembering the independence and unity of purpose that farm life had imparted to my grandfather and his siblings. Fifteen years before, prior to my grandfather's death, there had still been five of them, pushing into their late eighties and nineties with a common jollity.

The drive to Cleveland from Hale Farm condenses the historic American journey from farm to industrial city and from city out to suburb. The whole story is there. My rural reverie faded as the houses and subdivisions grew denser. The suburbs, as my husband likes to remind me, usually with a smirk, are full of people like me who long for the country. And as more of us move there, we whittle it away.

Left behind was not only the Hales' former farm but the ghosts of their erstwhile neighbors—from thousands of farms that no longer exist. Between Cleveland's heyday, around 1920, when its population was nearing a million, and 1960, when its population was draining into the suburbs, thirty-three hundred farms disappeared from Cuyahoga County—leaving just over one hundred small farms by 1960. The average acre of Cuyahoga County land in 1960 sold for thirteen thousand dollars, a price conducive to small suburban lots.

The road from Jonathan Hale's farm leads to Cleveland, and the story of Cleveland's growth is one of the most remarkable stories of nineteenth-century America. While the frontier moved west, shaping American society through an increasingly speculative approach to land use, the great American industrial cities rose and attracted population—a crucible in which the relationship of Americans to their resources forever changed.

In Cleveland, the relentless pursuit, transport, and use of oil, iron, and other resources employed the masses. The city worker's standard of living depended not on his own industry but on what was left after subtracting his rent from his wages—both factors in the equation of other mens' profits. This equation kept workers' wages low and rents and other expenses high, and it accounted for the spreading slums in cities such as New York, Boston, Cleveland, Detroit, and Chicago. The individual's access to resources—the ability to farm, garden, fish, hunt, keep a cow, a goat, a few chickens, an apple tree, berry bushes, or tomato plants—succumbed to space constraints, poisoned rivers and soils, and health and zoning regulations protective of sanitation.[28]

In Cleveland, a proud, progressive Midwestern optimism rested on the industrial innovation fueled by abundant oil, steel, and enterprise. A sense of civic commitment led to the building of cultural showpieces and to weak attempts at social reforms. But determined optimism and ruthless productivity, so cherished by the leaders of the young industrial cities of the Midwest in this period, not only failed to counter the forces of social and economic displacement at work in this period but also failed to acknowledge them. And in so doing, this mind-set allowed thousands of newcomers to the city to find their fate in the slums that grew up to defeat the city's promise.

CHAPTER 8

CLEVELAND: THE INDUSTRIAL CITY

As long as you have a boundless extent of fertile and unoccupied land, your labouring population will be far more at ease than the labouring population of the old world; and, while that is the case, the Jeffersonian polity may continue to exist without causing any fatal calamity. But the time will come when New England will be as thickly peopled as old England. Wages will be as low, and will fluctuate as much with you as with us. You will have your Manchesters and Birminghams; and, in those Manchesters and Birminghams, hundreds of thousands of artisans will assuredly be sometimes out of work. Then your institutions will be fairly brought to the test.

—Lord Thomas Babington Macaulay,
letter to Henry Stephens Randall, 1857

The weekday bustle of present-day downtown Cleveland gives way to quieter weekends. In the office and government buildings that line the great avenues, many stores and eateries close for the weekend, and fewer pedestrians share the sidewalks. The white office workers have gone home for the weekend and taken their spending power with them.

Among the silent buildings, the Public Auditorium occupies what was once an entire city block. Completed in 1928, its cornice bears the proud legend "A Monument Conceived as a Tribute to the Ideals of Cleveland Builded by her Citizens and Dedicated to Social Progress Industrial Achievement and Civic Interest." Beyond the auditorium, the massive Cuyahoga County Courthouse and its twin, the "new" City Hall, stand like the great bookends of a monumental building project. Near the auditorium, the education administration building looms over a large, dry lawn, and one block farther into town, away from Lake Erie, are two more human-scaled but dignified library buildings.

The neoclassical optimism of the aging buildings seems quaint in the modern cityscape. Like the proud works of Shelley's Ozymandias, crumbling in the desert, they seem to have outlived the audience they were meant to impress—their origin and meaning a mystery. The incongruity begs the question not only of what happened, as in Shelley's poem, but of whether the myopic focus of the monuments' creators in some way conjured the surrounding lifelessness.

These are the buildings of the Group Plan: one of Cleveland's greatest civic undertakings, hailed around the nation in the early twentieth century. The buildings, once connected by malls designed for formal gardens, still assert the virtue and aspirations of a bygone citizenry. The marble and granite facades are the legacy of the industrial and urban boom that even Jonathan Hale's children had felt, halfway down the Cuyahoga Valley to Akron, as Cleveland's growing population created a market for agricultural products throughout the region, and tired city residents sojourned to the Hale Farm Inn for refreshing weekends in the country.

Through a growing web of railroads moving freight and passengers, nineteenth-century Cleveland gradually transformed the region around it. These changes continued in the twentieth century as Cleveland's population dispersed into the surrounding countryside, as in many twentieth-century American metropolitan areas. Cleveland's core of proud government buildings became the setting of a long decline from which Cleveland has yet to completely rebound.

At its zenith, just before the Depression, when the first buildings of the Group Plan were completed, Cleveland held almost one million people and was the fifth largest city in the nation after New York, Chicago, Los Angeles, and Philadelphia. It was the age of the industrial city: two other great manufacturing cities of the era, Detroit and Pittsburgh, had metropolitan populations comparable to Cleveland's. All had surpassed former transportation hubs such as Buffalo, which had lost its canal-born prominence to the railroads, and St. Louis, former gateway to the West.

America's age of manufacturing is now past. The shift to a service economy and then to an information economy, the new communication and information technologies, the trend of manufacturers relocating overseas to take advantage of cheap labor and tax breaks—all of these changes have left the great industrial cities of America's past high and dry. Cleveland's current population is roughly half that of its prime. It now ranks twenty-third among U.S. cities in terms of population. On the other hand, its greater metropolitan area has sprawled to meet Akron's and has added a couple of million people during the twentieth century. The Cleveland-Akron metropolitan area ranks as the sixteenth most populous in the nation.

Cleveland's quiet avenues and bustling industrial past invited questions about the Americans who had followed Jonathan Hale—their communities, aspirations, and resources. Before heading to the library to look for clues, I stopped in a drugstore for a pad of paper. The young cashier, like an oracle of the deserted district, pronounced, in a few words, the epitaph scrawled by the modern city on the neoclassical repose of the Group Plan.

"Things are more expensive here than in Boston," I remarked when my purchases came to an unfamiliar total.

"We got a high sales tax," she replied.

"You do have a high sales tax," I agreed. "I noticed that. What is it?"

"Eight percent," she deadpanned.

"Eight percent. Wow."

"It all go right to the school superintendent."

"Does it? And how are your schools?" I asked.

"Crummy."

VICTORIAN CLEVELAND

Cleveland's schools had been in the paper the previous morning. The district held 120 school buildings. The tax—an increase estimated to yield $335 million—supported a "massive reconstruction project" begun in 2001 after the roof of a school gym collapsed.[1] Despite Cleveland's dreams of social progress, touted by the auditorium cornice, the city's schools struggled with the problems that have faced so many American city schools since the twentieth-century exodus of the middle class to the suburbs. If social progress had brought Cleveland to have the same poor urban schools and ghettoized black population as so many other American cities, what exactly had the "social progress" heralded on the auditorium's cornice meant to Victorian Clevelanders?

At the dawn of the twentieth century, when the Group Plan was envisioned, Cleveland emblemized the hopeful boosterism characteristic of Ohio and other cities and towns founded in the glow of young nationhood. Local officials and business leaders shared Emerson's faith in a link between commercial and social progress—or perhaps Emerson shared theirs. Local government was an amateur profession shared by the business leaders serving on various committees and making philanthropic donations where called to do so. Local pride and investment served to attract further investment, including the important rail and transportation connections that could make a town a regional hub or bypass a town and leave it to wither.

In this, Cleveland was very successful. In the second half of the nineteenth century, Cleveland gradually usurped the primacy of Cincinnati, the "Queen City of the West." With Cleveland's location on Lake Erie, its city fathers ably translated the wealth generated by gritty factories and refineries, mechanical innovation, and trade with other mushrooming Great Lakes cities such as Chicago and Detroit into civic institutions and other symbols of cultural achievement.

When Cleveland's citizens raised a statue of Commodore Oliver Perry, hero of the Battle of Lake Erie in the War of 1812, in their Public Square in 1860, it was the first memorial in Ohio, a state whose largest cities now sought to guild commercial success and civic pride with eastern-style refinements. Only thirty-five years later, in the aftermath of the famous "White City" of the 1893 Columbian Exposition of Chicago, the Cleveland chapter of the American Institute of Architects was recommending a plan to "group" the new government buildings needed by the city in a Beaux-Arts scheme that—as developed by the author of the White City himself, Daniel Burnham, or rather his firm—would draw national attention to Cleveland.

When Burnham's firm submitted its drawings in a report to the mayor, the report noted that "The drawings were awarded a gold medal at the World's Fair in St. Louis and have been exhibited at the Architectural League of New York, the Architecture Club of Chicago, and at the Museum of Art in Toledo." The report's flattery continued:

"Cleveland is today being taken as an example of what civic pride may accomplish." When the plans were unveiled in Cleveland in 1903, President Roosevelt attended.

However, darker forces of urban growth pressed upon Cleveland, its leaders and inhabitants. The pressures of rapid growth, industrialization, and immigration that swelled the ranks of the poor far outpaced social investment, philanthropic efforts, socioeconomic theory, and the reform and modernization of local government that was required before government could administer massive new urban problems. These powerful forces shaped the city. The Group Plan buildings emblemize the city fathers' determination to beat back the bedeviling pressures of urban transformation. Yet the buildings also emphasize the naïveté of their strategy, the inadequacy of their weapons. But it would be wrong to look condescendingly at their efforts; even today growth, poverty, and technological change continue to have, in many ways, the last word in urban design, as policy struggles to catch up.

In the words of historian Melanie Simo, reform-minded Americans of the Victorian era wanted "social reform without social upheaval" or "a genteel social responsibility." Victorian Cleveland was, indeed, reform minded but constrained by the magnitude of the social problems it faced as well as by recent history. Real understanding, generosity, and empathy existed, as shown by the arrival of the neighborhood, but in very small doses. But enormous numbers of the needy lay beyond the reach of such limited private measures.

The abolitionist movement, after all, had culminated in not only the liberation of enslaved African Americans but also the national devastation of the Civil War. The long period of postwar rebuilding focused increasingly on the accumulation of wealth. Although the labor movement and the women's rights movement gained momentum in this period. Beautification projects symbolized progress in ways that tended to unite a populace, whether in the tree-planting projects of ladies' auxiliaries or the grander schemes of the City Beautiful Movement. The City Beautiful Movement, of which the Group Plan was one expression, enshrined the American longing for a noble and culturally polished national identity. And it diverted attention from the intractable problems of the poor.

The Group Plan was an outgrowth of the City Beautiful Movement, a movement rooted in the Victorian desire to guide urban transformation in genteel ways. The grand plans of the City Beautiful Movement, implemented in many American cities in the late nineteenth and early twentieth centuries, asserted the idealization of progress over its sooty reality with monumental facades. By refining both the appearance and cultural life of a city, the schemes of the movement envisioned cities that would rise above poverty through their sheer incongruity with it. "Beautiful and clean cities attract desirable citizens, and real estate values increase," advised landscape architect Loring Underwood, lecturing the townsfolk of Concord, New Hampshire, in 1909.[2]

The City Beautiful Movement—with its benefits of inspiration and unifying beauty—conferred more concrete benefits, too. It provided parks, malls, and grand public spaces for the citizen's use, even the poor citizen. Landscape architect Frederick Law Olmsted deserves special credit for merging the goals of beautification and social relief through the provision of large parks and park systems in the Victorian cities of America. These "lungs" of the city, as he called them, created healthful recreational settings that, he rightly insisted, would serve all classes and would allow them to mingle, an idea that the *New York Herald* greeted as a scandal during the construction of Central Park.[3]

The City Beautiful Movement also expressed a growing American faith in city planning, as a corollary to the organizational triumphs of business and another manifestation of human progress and prowess. Captains of industry from Massachusetts to Chicago inspired some of this faith as they attempted to create model worker communities throughout the nineteenth century to house reliable labor free of the vices rampant in slums.

George Pullman explained to an eager press corps the landscaped cluster of buildings he had built at Pullman, Illinois, outside Chicago, to house the employees who manufactured his famous railroad cars:

> I have faith in the educational and refining influences of beauty and beautiful, harmonious surroundings, and will hesitate at no expenditure to secure them.

Built in 1881, with garden apartments, a church, theater, hotel, and library, Pullman remained a paternalistic fantasy, as did other company towns, doomed by the unchecked self-interest of their owners. The various small congregations in Pullman, for example, never had enough money to rent Pullman's church. And when the railroad depression of 1893 caused Pullman to lower wages, he failed to lower rents, too, and havoc ensued. In 1898, a year after George Pullman's death, the Illinois Supreme Court essentially outlawed company towns, in which individuals could exercise complete control over the lives and economic condition of their workers.

Victorian Cleveland had been a city of proud stone buildings of a more human scale than the existing auditorium and its kin. The city boasted solid schools, four-story palaces of commerce at major intersections, numerous hotels, several churches in each residential ward, two "orphan asylums," a "charity hospital," a jail, a prison, a police station, and a growing number of factories and refineries along the Cuyahoga River, from which webs of train tracks and plumes of chemicals fanned out, as they did also from the wharves on Lake Erie.

Cleveland had become, by then, an industrial capital producing in its factories, workshops, and refineries the very nuts and bolts of the Industrial Revolution—all that was needed for manufacture and the production of wealth. John D. Rockefeller's Standard Oil Company sprawled along the Cuyahoga beside the Cleveland Paper Company. A number of riverfront ironworks produced machinery and parts of various kinds. All around, in warrens of steam-powered workshops, inventors and their helpers experimented in improvements to electricity, motors, batteries, and springs and produced every kind of mechanical device that could further or refine industrial processes. Small businesses that began in these workshops often grew to be large industrial concerns, making wealthy men of their tinkerer-founders.

By 1881, the industrial heart of the city included a gas and coke company, Otis Elevator, a manufacturer of "agricultural implements," a lubricating oil works, bridge works, railroad car works, nut and bolt manufacturers, wheel makers, assorted forges and foundries, an engine factory, a machine shop, a sewing machine factory, brass and pipe works, paint and color works, two barrel works, a hardware manufacturer and shovel maker, a varnish factory, a brewery, a malt house, a box factory, a sawmill, and a planing mill for lumber. A slaughterhouse stood near related tanneries and candle and lard works. Downtown, five banks lined Superior Avenue, and the Cleveland *Plain Dealer*'s headquarters stood nearby along with hotels and commercial buildings. As from the beginning, the Public Square with its heroic Civil War memorial declared the heart of the city.

By 1884, Cleveland had its first electric trolley line, serving wealthy Euclid Avenue, also known as "Millionaire's Row." Rockefeller and other grandees built stately mansions heading eastward out of town and away from what one contemporary observer described as the "blackness, dirt and decay [that] were visible everywhere."[4] At the end of the Euclid Avenue trolley line, the city's wealthy industrialists began to create a new park-like cultural district in their own enclave away from the heart of the city. Jeptha Wade began the trend by giving some of his land for a city park, adjacent to which the Cleveland Museum of Art was eventually sited with support from his grandson. The colleges that would eventually become Case Western Reserve University sited here as well, and with the arrival of additional cultural institutions, the entire area became a green and prosperous oasis known as University Circle. But the stunning level of cultural investment in the area remains somewhat secluded. The downtown, on the other hand, has been refitted with modern sports stadia but remains somewhat culturally bereft.

The earliest insurance maps of Cleveland give few clues about how industrial workers lived. Several enclaves near the industrial sites were subdivided into tiny lots about twenty-five by fifty feet—large enough for very small homes, crowded together. Sometimes factory owners, bankers, utility owners, or other men of wealth owned worker housing on which they collected rents. In 1900, certain wards contained mainly rental housing, providing clues to workers' living conditions. In those wards, some near the industrial heart of the city, the population density was thirty to fifty people per acre, generally without running water. Death rates in Cleveland were slightly higher than average for the state, typical of cities of that era. In certain, more crowded wards full of renters, the death rate ran higher still.[5]

GROWTH AND GROWING CONTRADICTIONS

Between 1890 and 1910, Cleveland's population would more than double to over half a million people, due in large part to immigration. Between 1900 and 1914, approximately five thousand people, mainly Europeans, arrived at Ellis Island every day. Many made their way up the Hudson and westward to disperse via the Great Lakes to the mighty industrial cities of Cleveland, Detroit, and vast Chicago. The immigrants who poured into Cleveland and other northern American cities came in waves—a human tide that easily overswept the fragile urban systems of the Victorian cities and Victorian customs of local government. During some of those years, three-quarters of Cleveland's population was "either foreign-born or of foreign parentage."[6] The city experienced the terrible strains of rapid growth, with "roads, sewer system, and health and housing codes [that] were more characteristic of the mercantile town it had been."[7] In 1910, in one ward of 7,728 people, a survey of living conditions found only eighty-three bathtubs.[8]

An industrial and commercial profusion attracted this human profusion—offering both jobs and wares to purchase. Insurance maps of the period show the early commercial and material exuberance that would overtake America as the twentieth century progressed. No longer just a producer of the raw materials and mechanical parts needed for industrial development, Cleveland now offered a wealth of consumer goods, services, and entertainments that enticed people to dispose of their new wealth. Laundries multiplied. A cloak factory, drug and chemical stores, bakeries, sandwich shops, peanut roasters, a butter manufacturer, and a crockery warehouse all appeared. There were more theaters, more sellers of candy and clothes, and, finally, a casket maker. Livery stables were everywhere. A bowling alley opened. Large shopping arcades appeared just before the turn of the century. Wealth and invention together promoted advances in building technologies and services—as shown by the existence of a new electric company and an automatic fire alarm and sprinkler factory. Production, invention, and investment polished the city's cultural life, lured further investment, and drew thousands of job seekers, to whom the luxuries around them would seem a mirage.

Thus, in the late nineteenth and early twentieth centuries, growing urban populations plus expanding industrial and commercial zones restructured the town-like layout of American cities into larger, more modern urban structures in which different uses occupied different zones[9] well before zoning codified this tradition. By the dawn of the twentieth century, intra-urban transit systems of horse-drawn omnibuses had given way to electric street car lines and early subway systems, which further enabled the growth of the city and its suburbs. Where once everyone had lived close to their work in a relatively small city such as Cleveland, streetcars took wealthier citizens from work to their own nicer enclaves. And the poor, who had once lived near or with the rich in order to serve them, had by this time become more concentrated in certain districts, giving rise to slums.[10]

Immigrants arriving in Cleveland in the late nineteenth century fared worse than earlier arrivals. While many early-nineteenth-century immigrants arriving in Cleveland had found their way to land and prosperity—old and prosperous German towns and neighborhoods attested to this—land was far more expensive and less available a century later. Low-paying industrial work under poor conditions came to be the immigrant's expectation. Nonetheless, foreign immigrants tended to form city neighborhoods that, though poor, featured systems of mutual aid, and residents could move outward into other areas if they prospered.

At this point, the fate of black Americans began to diverge from other new arrivals to the city. African Americans were kept together by prejudice that closed all housing markets to them outside the few traditionally black neighborhoods in cities such as Cleveland.

The beginnings of the black ghettos that would become characteristic of America's major cities can be seen at this time. Not only were the neighborhoods of black residents closer to noxious uses such as industry, but vice districts developed adjacent to the incipient ghetto partly because "the predominantly white police forces of cities, responding to the pressures of white public opinion, often refused to allow red-light districts to develop anywhere except in or near a black neighborhood."[11]

The 1910 maps of Cleveland that show so much new commercial activity also reveal the slums that had grown up as a failure of "genteel social responsibility." Near the heart of town, on the future site of the Group Plan, sprawled tenements, settlement houses, and charities. It was one of the city's black enclaves, described by Kenneth Kusmer in *A Ghetto Takes Shape:* "a lower-class neighborhood of dilapidated structures, stretched along Hamilton Avenue between East Ninth and East Fourteenth streets. Close to the industrial section and the docks, it was the least desirable neighborhood for blacks to live in." And therein lies the origin of the Group Plan. The plan, it turns out, had been an early slum clearance project.

Even as Victorian sentiments and the City Beautiful Movement gave way in some cities—such as Cleveland—to a more modern twentieth-century Progressivism, schemes such as the Group Plan held the attention of the public and political leaders. The Group Plan received the resources of government. The poor did not. And as government failed to understand or solve the knottier problems of urban transformation, more and more people who could afford to do so left the city. Suburbanization—a slow-motion flight of the well-to-do to more salubrious surroundings—transformed the city and its region still further. Suburbanization was the spontaneous, market-bound response of the populace to urban ills—or, rather, of the part of the populace that could afford the solution. It transformed the city, the American landscape, the American economy, and American government. And it gained enormous momentum both from the tremendous growth of the late-nineteenth-century city and the immigration and industrialization behind that growth and from the refinement of transportation technologies to carry people between city and suburb. But we cannot move to discuss suburbanization without considering the industrial city through the eyes of a Victorian named Henry George, who found the sources of growing industrial wealth and poverty in the American system of land distribution. George's theory attracted the attention of Leo Tolstoy and Albert Einstein, and when George came to Cleveland to lecture, he found disciples there.

CHAPTER 9

TOM JOHNSON AND HENRY GEORGE

Equality of political rights will not compensate for the equal right to the bounty of nature. Political liberty, when the equal right to land is denied, becomes, as population increases and invention goes on, merely the liberty to compete for employment at starvation wages.

—Henry George, *Progress and Poverty*

The fight to transform Cleveland's Victorian government to a twentieth-century model fell to Tom L. Johnson, the only mayor of Cleveland to have his statue placed in the Public Square. As mayor from 1901 to 1909, he devoted himself to reform. He fought to create a public street car system with affordable fares—and to modernize and professionalize city government, taking it back from "privileged" influences and equipping it to confront the problems of the twentieth-century city.

Johnson's election was part of a Progressive renaissance that briefly established itself in the proud, earnest Midwest as that region grappled with the social problems of the industrial city. During this period in the early twentieth century Cleveland elected two Progressive mayors, including Johnson, while Detroit and Toledo, Ohio, each elected one.

Johnson's idealism found sustenance in the writings of a more famous reformer and philosopher of the period, Henry George. Johnson became a lifelong friend and follower of Henry George and a devotee of George's theories on industrial progress and poverty. Tom Johnson was not the only Cleveland business leader who found his avocation in pursuing George's theories. John C. Lincoln gave part of his fortune to found the Lincoln Land Institute for the study of land use policy. But Johnson's close friendship with George, who urged him into public life, lasted until George's death and became a collegial alliance. Johnson, an uneasy public speaker as a young man, sometimes found himself standing in for his mentor. Johnson and George, who eventually became summer neighbors in New York, shared a political idealism that focused on the problems of the industrial city.

Henry George was a born observer and a keen one. He observed American life for three and a half decades, usually as a journalist scrapping for income, before he began to translate his observations into his masterwork, *Progress and Poverty*, published in 1879. He was forty years old and trying to feed a young family as a printer in San Francisco when the book came out and made his reputation. He had become an economic philosopher. George's response to growing nineteenth-century poverty—to inquire into and observe the systemic causes—was startlingly rare. This aspect of his work alone was noteworthy and harkened back to a time when political thought, the thoughts of America's founders, held a greater and simpler profundity.

George, who praised private property as an incentive to work, was no Marxist. (Marx had not yet been translated into English when George wrote.) But he railed against private landownership as promoting a monopoly on the resources that people needed in order to work, to be productive, and to earn a living.

HENRY GEORGE AND INDUSTRIAL AMERICA

When we speak of industrialization in America, we vaguely acknowledge a process by which the majority of breadwinners moved from an agricultural to an industry-based livelihood. Henry George examined industrial America and the process described in previous chapters by which the common man's and woman's access to resources gradually diminished in nineteenth-century America. His examination of the relationship between industrial progress and poverty gained worldwide attention. In it, he explained why people without access to resources so readily became paupers. He saw that whoever controlled a resource, such as iron for steel, for example, controlled the livelihoods of everyone who depended on that resource to be productive and to earn a living.

George gave the example of one man owning an island where a hundred others live. That man, he pointed out, would have complete control over wages *and* rents and the ability to keep the hundred others in abject poverty—just like the industrialist who owns his workers' homes or the modern sugar plantation owner or the companies that own stores or housing in mining communities. Though George was careful to praise entrepreneurship and to avoid blaming capitalists for worker poverty, he blamed landowners—particularly speculators—for holding needed resources. And he castigated industrial "monopolists" who stifled commerce in their own interests. His thinking on monopolies animates the Sherman Anti-Trust Act of 1890, written by an Ohio senator and supported by Tom Johnson during his brief tenure as a member of Congress. And it seems likely that George's ideas influenced the U.S. Supreme Court when they outlawed company towns in the 1890s.

George saw that, as industrialization and urbanization increased together, the rapid rise in land values around cities tied up more and more land and resources in speculation just as they were increasingly needed to make the growing urban population productive. George never mentions the need to house the growing population and the increasing amounts of land required to do so, although much land speculation aimed, then as now, at the housing market. This effect is easier to see now as the U.S. population approaches three hundred million and home building begins to trump other kinds of economic activity in the race for land.[1]

George's ideas owed much to John Stuart Mill's *Principles of Political Economy*, published in 1848. *Progress and Poverty* reiterated Mill's observation that "[t]he essential principle of property" was to assure that people get to keep whatever their work or thrift permits them to acquire. "This principle," said Mill, "cannot apply to what is not the produce of labor, the raw material of the earth."[2] In this conclusion, Mill, like George, echoed Jefferson and a strain of eighteenth-century French economic thought. As Mill argued, "When the 'sacredness of property' is talked of, it should always be remembered, that any such sacredness does not belong in the same degree to landed property. No man made the land. It is the original inheritance of the whole species."[3]

George saw that private ownership of land had seemed sensible enough to the colonists when land abounded—although he noted the New England view of land as a community asset, assigned at the community's discretion. And Mill had provided this optimistic interpretation of the laborer in a young nation experiencing frontier conditions: "To begin as hired labourers, then after a few years to work on their own account, and finally employ others, is the normal condition of labourers in a new country, rapidly increasing in wealth and population, like America or Australia."[4]

But as the country matured and land grew scarcer and was increasingly engrossed by the wealthy, George saw the noose tightening on the working class. George did not suggest a redistribution of property but a tax on the profits to landowners that would equalize the effects of landownership. George's thinking was still echoing in political thought when, in 1919, Benton MacKaye declared that land "is the source of all employment," but MacKaye dropped the idea of access to resources as a natural right, an idea increasingly unfamiliar to the American public.

Surrounded by the American habit of land speculation, George went into far greater detail than Mill on the harmfulness of private landownership, and particularly land speculation, in holding resources for production off the market, hamstringing production, and causing periodic shortages and price spikes even as industrial progress boomed. George's theories helped to explain the slightly larger fortunes of those who controlled resources during this period versus those industrialists who built their fortunes on mechanical refinements: John D. Rockefeller with his oil and Andrew Carnegie with his steel were examples of resource monopolists. His theories explain the monopolistic effects of modern industries that expand vertically to harvest the raw materials they use.

George's downfall came not from his critique of the system—although some found his philosophical conclusions about growing inequality distasteful—but from his suggestion that something be done about it. His attempt to write a specific prescription for the ills he observed, in the form of a new tax code, earned him scorn from his critics, mainly those whose financial interests his ideas threatened.

Still, George's tax reform ideas find their modern manifestations in all taxes that emphasize the value of land versus the improvements made to it; in surtaxes on absentee ownership; and in high real estate taxes to discourage speculation, though these are all relatively rare.[5] As mentioned earlier some cities and counties successfully use a very small tax on real estate transactions to underwrite the purchase of land for public use—generally for recreational enjoyment but sometimes housing. Some cities adapt George's concept of taxing the value of land to encourage redevelopment. Economists Wallace Oates and Robert Schwab credit the land-value tax for the recent partial renaissance of Pittsburgh, whose comeback has to date outdistanced Cleveland's.[6]

George's voice, urging a reconsideration of the governance of the distribution of resources, was like a last, lonely echo of the founders. George also described, as no else seemed able, the effects on resource distribution as the United States converted from an agricultural to industrial economy. Only Jefferson had ever appeared to foresee the new policies regarding land use that would be necessary as the nation industrialized. In his famous 1785 letter to Madison he remarked presciently, "If for the encouragement of industry we allow it [land] to be appropriated, we must take care that other employment be provided to those excluded from the appropriation."

George's treatise was the most widely read writing on economics of his day, probably owing partly to both his lucid style and his gripping idea. His conclusions attracted praise from the great thinkers of his day—Tolstoy, Dewey, and later Einstein—as well as from many less well-known citizens of the day who fought for social reforms, one of whom was Tom Johnson.

Tom Johnson was born fifteen years later than Henry George. He was six when the Civil War broke out, separating his family from their wealth. They retained, however, their top-drawer connections: Johnson began his career keeping books for a street railway company owned by a family acquaintance. At the age of twenty-two, he invented the glass fare box for street railways—the kind still used on buses and subways today. That invention earned him a small fortune, which he translated to the purchase of his first street rail company.

Johnson was still young, the wealthy owner of a number of street railways and steel manufactories that produced the rails they ran on, and a self-confessed monopolist, when he had a near-religious conversion to George's ideas after reading *Progress and Poverty* and another tract of George's. He met George in New York in 1885 and befriended him. He bought land for George's family alongside his own vacation home in

Fort Hamilton, overlooking the entrance to New York Harbor. His political awakening and George's urging led him to run for Congress, where he supported the antitrust legislation of 1890 and 1892.

George and Johnson sometimes traveled together, and Johnson sometimes accompanied George on his speaking engagements, during which George encouraged Johnson to practice his public speaking by answering the audience's questions afterward. George sometimes stayed at Johnson's home on Euclid Avenue in Cleveland. Johnson gave time and money to George's first run for mayor of New York, on behalf of labor groups, and ran George's second unsuccessful campaign for mayor of New York in 1897. Johnson wept at George's deathbed that same year.

After George's death, Johnson gradually gave up his lucrative businesses. He took the Progressive fight into the trenches. He served as mayor of Cleveland from 1901 to 1909, where he used up his wealth and health fighting the advantages of "privilege" that had once served him well and attempting to help the common man by promoting reforms of government. When he died in 1911, having made many powerful enemies but also having earned a place as Cleveland's most beloved mayor, he was buried in New York in the Johnson family plot by George's grave.

Johnson fought for home rule—that is, local autonomy from state supervision—to allow local governments flexibility in addressing the new and varied problems facing growing municipalities. He fought to place professionals in charge of different departments of government—rather than the well connected but unskilled. And his biggest fight was for municipal ownership of utilities. Having spent a good portion of his early business life seeking and obtaining city grants for street car lines, Johnson well understood the corruption this system brought to government. Wealthy utility owners did favors for politicians and sought them in return. These favors extended well beyond obtaining business from the city. The wealthiest residents promoted candidates and influenced policies regarding taxation, labor relations, and whatever might affect the growth of their own wealth or business. Johnson fought this system of influence, dedicating himself especially to one symbolic effort: the construction of a city-owned street railway that would convey Cleveland's citizens to their work and homes across the ever-growing city for only a three-cent fare. The proponents of "privilege," as he called it, fought him every step of the way in endless court battles. Johnson accused them of "government by injunction."[7] Though construction on the railway was begun, the state courts eventually ruled that the city could not own and operate a railway.

Though Johnson lost the battle, many of the government reforms he fought for occurred after his death. Government began to incorporate more professional expertise in its management functions. Municipalities in most states have gradually gained greater autonomy. And reforms in municipal utilities occurred. Government did modernize, though always lagging behind the problems it modernized to solve. For example, despite the obvious pressures on existing housing stock created by industrialization and immigration, America would not consider the question of how to house the poor that overwhelmed cities such as Cleveland until Franklin Roosevelt took office in 1933. However, Johnson's three-cent fare partly addressed the worker's need for greater flexibility in finding accommodations by allowing him or her a greater geographical range in which to find a home.

CITY UNDER SIEGE

By the end of Tom Johnson's tenure as mayor, the problems of the American city had just begun. The continued immigration of large numbers of poor, the ascendancy of the car, and national economic trends would hollow out cities throughout the twentieth century, redistributing population and creating the vast metropolitan areas of today.

Between 1910 and 1920, when much of the demolition for the Group Plan occurred, the continuing migration of those falling off the ladder of the agricultural economy swelled city populations further. The Great Migration of African Americans, who abandoned rural poverty in the South to seek jobs in the cities of the North in large numbers during this period, tripled Cleveland's black population, bringing it to over thirty-four thousand, mainly confined by the conventions of prejudice to the existing black neighborhoods of the city. In nearby Detroit, the black population increased over 600 percent. And in Chicago, home to the largest black population in the Midwest, the number of African American citizens more than doubled.

Though African Americans in Cleveland had initially been more dispersed throughout the city, segregationist sentiments grew with the black population.[8] In addition to concentrating the black population in three main areas of the central city, this growing hostility caused the death of black businesses that had once served whites, such as restaurants, caterers, barber shops, and clothiers. Increasing job and housing discrimination pushed blacks into greater poverty and caused an increase in crowding and a decrease in housing investment in these areas, writing another chapter in the history of the American city—one of disinvestment and abandonment.

As the twentieth century progressed, urban growth created larger and larger versions of this same recipe for failure in cities: growing zones of business, noxious industries, poverty, racial ghettos, and vice districts. This fomented another kind of "great migration." The flight of affluent whites to the suburbs that had begun with the transfer of the wealthy away from downtowns in the nineteenth century now gained momentum. Together with the automobile, the trend toward suburbanization would remake American settlement patterns and create a new suburban culture of escape in which the notion of civic responsibility was too easily limited to regulating lawn aesthetics through neighborhood associations and the endless fight for a decent public education for the children, generally denied the poor.

American land had once seemed a solution for social problems—giving the poor a leg up. It now served to help many Americans escape social problems without solving them. But the solution of the Founding Fathers—to allow poorer Americans to emigrate to the abundant lands near the frontier—had always contained an element of escaping rather than confronting the problem. The founders had allowed natural abundance to provide for the poor. With natural abundance diminishing, the poor—expanding in numbers if not percentage of the population—had to rely on the less bountiful gifts of their countrymen.

The ideas of George and Johnson, while distantly echoing those of the founders, are contemporaneous to a society very different than that of the founders. In their more crowded society, greater limitations to a citizen's access to land and resources became acceptable. The questions raised by men such as George and Johnson asked whether the redistribution of resources brought by growth and industrialization would be accomplished by government design or by economic forces. Economic forces won. They continued to shape American cities, the cities that, in the present day, stretch from horizon to horizon.

The problems of the modern city continue to outmatch and transform government, just as the transformation of the nineteenth-century city once outmatched the city fathers of Cleveland and led to the administrative modernization initiated by Tom Johnson. Just as Victorian local government had to become a twentieth-century government, our current governments are evolving and giving rise to more and more regional authorities to deal with growth. But, as in the past, solutions lag well behind the difficulties they address. Government officials fail to see or articulate the larger trends closing in on us. And now, as in the past, the public will to act seems mobilized only by the direst circumstances, rarely in advance of disaster.

CHAPTER 10

A HOME OF ONE'S OWN

Poverty deepens as wealth increases, and wages are forced down while productive power grows, because land, which is the source of all wealth and the field of all labor, is monopolized.

—Henry George, *Progress and Poverty*

As cities grew and the agricultural economy continued to yield to industrialization, the American government of the young twentieth century turned its attention from the need to provide its citizens with land to the need to house them. In the 1930s, the dire situation of Americans displaced by the Depression—exacerbating a prior shortage of housing for the not so well-off—helped to bring about the first attempts at a national housing policy.

In crafting a new housing policy, the federal government not only reevaluated the needs of its citizens but enlarged its subsidy of private industry in order to create new housing. In a gesture comparable only to the extravagant federal land grants thrown to railroads in the nineteenth century, the government sponsored new subsidies of the real estate and building industries. This policy shift reduced the government's role in distributing resources to Americans and invested additional power to apportion resources in industries and the market.

These industries had gained strength in the high-flying twenties, finding in Herbert Hoover, as secretary of commerce and then president, a receptive ally.[1] The Depression, however, curtailed Hoover's experiment with trickle-down economics. The Depression eased the chronic public distaste for poverty that underlay the trickle-down theory by pushing many more Americans into the ranks of the poor.

With so many Americans in need and suddenly sensitized to the federal government's potential to solve social and economic problems, Franklin Roosevelt was elected in 1933. He took the helm of a nation of people sorely in need of jobs and homes, including two million jobless construction workers. And to solve both problems, Roosevelt crafted a larger role for the government in providing housing. He put jobless construction workers on the federal payroll to build model communities that would both house citizens and serve as models. I was born in one—Greenbelt, Maryland—while my father finished college and then earned his master's degree on the GI bill in the mid-1950s. Although private industry sued the government for building these communities—and thereby reoriented the government's approach to one of subsidy to those industries—these communities were not just the work of the federal government but a natural result of increased public concern over housing and the work of earnest reformers.

During the first decades of the twentieth century, local governments helped draw national attention to the need for decent, affordable housing. As Benton MacKaye pointed out, the real value of wages fell in the first decades of the twentieth century. The gap between rich and poor had increased again. World War I helped to correct this imbalance, and home building boomed in the twenties, but the new homes were built mainly for those in the top third of the income spectrum.[2]

In cities such as Cleveland, the cancerous nature of slums and the need to house waves of immigrants and industrial workers drew reformers and engendered various experiments in sheltering and aiding the poor well ahead of any government intervention. Private individuals and groups founded five settlement houses in Cleveland in the decade around 1900, inspired by experiments in New York and abroad, and a strain of Protestantism that embraced social reform.[3] Rather than providing food and shelter, early settlement houses provided a caring staff that lived in the neighborhood and provided education and assistance for neighborhood residents of all ages and recreational programs for children in particular. Activities at the highly regarded Goodrich House in Cleveland even gave rise to other agents of reform, such as the Consumers' League of Ohio and the Legal Aid Society.[4]

In 1904, the civic-minded Cleveland Chamber of Commerce sponsored a study that concluded that "poor housing causes a whole litany of social and moral evils."[5] Another study in 1917 sounded the alarm again and revealed conditions in the black slums so deplorable that the chamber funded a real estate brokerage to work with African Americans in the area, although it is unclear what effect this had as prejudice continued to close other markets to blacks.

In response to the first study, Mayor Tom Johnson instigated a reform of the building codes. And Johnson also championed the cause of home rule, arguing that the special problems of cities required that they have the freedom to innovate. But Mayor Johnson had pitched his greatest battles in other fields. Serious housing reform in Cleveland awaited another great Progressive.

In 1911, eleven-year-old Ernie Bohn arrived in Cleveland with his father from Rumania. His mother had recently died. He learned firsthand the difficulties faced by immigrants seeking work and housing. He had a lame foot but a quick mind, and as he made good, he kept those lessons of hardship with him. He graduated from the public technical high school, then from Case Western Reserve University, and finally from Western Reserve Law School. Providing the "deserving poor" and working class with decent housing became the great passion of his life. By his early thirties, he had begun to influence housing policy at the state and national levels.

Young Bohn came to public prominence in 1929, when he was elected to the Ohio House of Representatives. He then became a city councilman. In 1933, in the state capital of Columbus, Bohn sought and won passage of legislation that established Ohio's Public Housing Authority laws, the first in the nation. He then became the founding head of Cleveland's new Metropolitan Housing Authority, which he led until 1968.

One of Bohn's early efforts at Cleveland's housing authority was an analysis of the city's slums. He commissioned a locally famous study by the Reverend R. B. Navin, "An Analysis of a Slum Area in Cleveland," completed in 1934. Navin set out to show that the provision of decent affordable housing and the eradication of slums would be an act of shrewd self-interest on the part of any municipality because slums, in addition to being a hazard to its residents, are bottomless money pits for the community, with the minimal tax revenues gained from the slum offsetting only a tiny portion of the city's cost for administrative services for the area, such as police.

Reverend Navin pointed out how the high human cost of slums mingled with the financial loss to society. The slums, he asserted, had a way of erasing whatever investment the city made in its occupants. "Such studies have proved conclusively that the vicious environment of the slums, more than offsets and more than counterbalances all of the uplifting influences of education, social service and religion. Research has shown that such areas breed vice and crime."

Only three years before, another city report had also tried to stress the inherent self-interest in helping the poor, this time the poor of Cleveland's large black ghettos. The 1931 "Economic Survey of Housing in Districts of the City of Cleveland Occupied Largely by Colored People" stated in a progressive report:

> Negro labor forms an important part of the backbone of Cleveland's industrial life. In the wide variety of industries operated in this district, they are employed in varying capacities. In the basic iron and steel industries they constitute a larger percentage than in any other class of work they do. This has been true for a number of years, particularly since the World War. These workers are engaged in many skilled and semiskilled operations; but the bulk of them are unskilled workers. They have been the reserve for Cleveland's industrial expansion for several years.
>
> Because of the importance of Negro workers to Cleveland's industrial life, the question of adequate housing facilities in the congested districts where they live in such large numbers, is one which is very definitely related to their industrial efficiency.

The good intentions of this report are as obvious as the author's inability to confront the racism and other prejudices underlying the plight of both African American slum dwellers and the working poor in general. The majority of society did not question industry's apparent dependence on low-payed, low-skilled workers, and this general acceptance precluded any suggestion of education, training, or better wages—any of which might have helped to ameliorate the situation of those surveyed in the study. Nonetheless, by emphasizing the usefulness of the slum-dwelling workforce to the community, the author tried to apply the one strategy that consistently mobilizes public attention: self-interest.

Despite the foreignness of the concepts of a minimum wage or government-sponsored housing to the public of this era, a feeling that society must house its poorest citizens to protect its own interests gained currency. The federal government, responding to both the Great Depression and pressure from state and local governments, began to pass housing-related legislation in 1934, when Franklin Roosevelt's administration created the Federal Housing Administration (FHA).

In Cleveland, Ernie Bohn saw the need for national legislation on housing. He lobbied for the passage of the U.S. Housing Act of 1937 and for the passage of other housing-related bills in Columbus and Washington, D.C. Since that time, over forty additional pieces of federal legislation have addressed housing or neighborhood development. But the response of government came late. By the 1930s, slums had transformed not only their own residents but the pattern of American settlement itself.

The flight of the affluent from cities to suburbs that began in the late nineteenth century was followed by the flight of, eventually, almost anyone who could afford to move. Historians usually attribute this exodus to growing industry, pollution, and congestion, as well as entrenched ghettos. The trend effectively sorted population into the pattern we see today: metropolitan areas of dense suburbs sprawling outward from an urban core whose density the twentieth century drastically reduced.

So, in the first half of the twentieth century, while reformers sought housing policies, growth surged in all industries related to housing—timber, construction, real estate, development, and finance—and these industries thrived. The new home-building industry, rather than public policy, determined the shape of the new suburban developments. And the home-building and related industries began a warm courtship of government that continues to this day—as well as a transformation of the American landscape.

At first with subdivisions for the well-to-do and later with tract housing, Americans' need for homes became a source of wealth to the private sector. New subdivisions on cheaper land outside cities offered a partial solution in the face of crisis.

However, allowing the private sector to take the lead in providing this partial solution carried a high price. The problems of cities remained unsolved. The sprawling suburban landscape was born, its patterns guided not by careful planning but by speculative investment and salesmanship, as unconnected subdivisions sprang up peddling ready-packaged lifestyles to targeted income groups.

Between 1929 and 1937, both the private and public sectors came up with remarkable prototypes for model communities. And yet, the ideas they showcased were sidelined by the race for profit. The private City Housing Corporation built the community of Radburn, New Jersey, in 1929. Architects Clarence Stein and Henry Wright planned the community through an ingenious marriage of pragmatism and idealism. The community was planned in superblocks bounded by main roads. Row houses along straight cul-de-sacs led from the main roads to the interior of the block, which was parkland for the community. The cost savings in paving and the laying of utilities by Stein and Wright's economical use of land covered the cost of the parkland and the extensive pedestrian path system that crossed the main roads in underpasses or on overpasses.

The Great Crash of 1929 left the City Housing Corporation bankrupt, but the federal government borrowed Radburn's model to build its own idealized communities in the 1930s: the Greenbelt communities emphasized affordable living for average American families. Like Radburn, the Greenbelt communities emphasized green recreational spaces, separation of pedestrian and automobile systems through under- and overpasses, cooperative community governance, and density. My mother remembers walking to the "movie house" or the small group of stores in Greenbelt without ever having to cross a street.

The government long ago ceased to own Greenbelt, Maryland. Renters were buying their apartments by the time I was born there and establishing a cooperative management. But the community is sufficiently dense that it is a stop on the Washington Metrorail system. Had suburban design continued to build on the values of Radburn and Greenbelt, we might not now face the sprawling, placeless landscape before us. But, as in Tom Johnson's Cleveland, the private sector stepped in to protect its interests by suing the government for its visionary planning. And the rest, as they say, is history.

THE TRIUMPH OF SPECULATION

The transformation of the American economy between the Civil War and World War I was complex and multifaceted. With increased invention, increasing industrial refinement and capacity, diversifying technology, and increasing capital, businesses grew and multiplied. Large corporations increased before and after the turn of the century. They increased their political and economic advantages through aggressive litigation. Financial institutions increased in variety and number as they have done throughout our history and with each economic change. Just as commercial banks and marine insurers had grown up to support the shipping trade of colonial America, savings and loans had grown up for the use of average Americans saving for land or a home, and these evolved further as home building evolved.

The spare wealth of the early twentieth century that put so many consumer goods and luxuries on display in turn-of-the-century Cleveland was a national phenomenon that enhanced the speculative mood of the country. This new wealth could be invested in order to gain more wealth. Since more Americans now had the income to take vacations, real estate in potential resort communities became a popular investment, as we have seen on Cape Cod. The stock market was another.

New businesses and trades gave rise to new business and trade *associations* in the second half of the nineteenth century. By the early twentieth century, national associations representing every kind of business interest gave a powerful lobbying voice to those interests, a development that would ultimately shape American life as much as population growth and urban expansion. At the dawn of the century, Teddy Roosevelt railed against the land laws that sold public lands to timber and mining interests while other citizens still had need of the public lands for homesteads. But money talked louder.

The National Association of Real Estate Boards (NAREB), established in 1908, began to influence national policy as the need for homes grew with the population. Their first battle was to reduce property taxes, contravening all of Henry George's work with brochures carrying simple, effective slogans such as "Let's Have Fairer Taxes." George would have had landlords give over much of their rent as taxes, in recognition of the fact that land belonged to everyone in the community *and* that rent was unearned income gained because the landlord held land that other people needed to use. When the federal government implemented the new national income tax in 1913, NAREB convinced Congress that a landlord's rent should not be taxed at all.

The more progressive ideas of the late nineteenth and early twentieth centuries were left in the dust of the popular rush to speculate. A national desire to emulate the well-to-do, with their increasing and conspicuous prosperity, eclipsed the concepts of a fair distribution of natural resources put forward by America's founders, by early French economists, by John Stuart Mill, and by Henry George. Like so many related developments of the twentieth century, increased leniency in property taxes both abetted speculation and contributed to the ease of homeownership. By the twenty-first century, only a third of Americans would fail to benefit from the combined effects of these policies and remain outside the circle of homeownership, though not all homeowners would enjoy a high or even decent standard of living.

By the early 1920s, real estate speculation became more frenzied, presaging the coming wild speculation in the stock market. The speculator gained a seemingly permanent ascendancy over the settler as a cultural icon. The cozy relationship of real estate business interests and government warmed further. As documented in Dolores Hayden's *Building Suburbia*, those speculators who bought land to subdivide it for residential resale now increasingly built homes on the land before reselling it. The construction of whole suburban neighborhoods that ranged from lower-middle-class tract housing to more affluent, picturesque subdivisions gradually came to consume—along with other forms of development—roughly two million acres of land annually in the 1950s and 1960s.[6]

Hayden credits Hoover and NAREB with bringing about the mortgage interest tax credit for homeowners that benefits homeowners today and, in 1932, the Federal Home Loan Bank Act "to 'establish a credit reserve for mortgage lenders' (and thus encourage home lending), and . . . the Reconstruction Finance Corporation to issue bonds to banks to help them cover mortgages." According to Hayden, banks scandalously misused both programs.

The Depression and its Hoovervilles—encampments of the poor—swept Hoover from office. But Roosevelt's attempts to provide housing also led to a backlash, as described earlier. Powerful development industry interests opposed these efforts in court, including NAREB, the U.S. Chamber of Commerce, the National Association of Retail Lumber Dealers, and the U.S. Building and Loan League.[7] These private interests sued the federal government, and the courts supported them, ruling that states could enable local agencies to take land, and sponsor slum clearance and the construction of low-income housing but that government must leave such projects to private enterprise.

These decisions helped formulate the momentous but compromised 1937 Housing Act for which Ernie Bohn had lobbied. Through it, the new FHA helped fund the low-income housing initiatives of local governments. The Housing Act made a number of important concessions to private developers that reduced its efficacy as an aid to the poor. The act required that every new housing unit built under low-income housing programs replace an older, demolished one.[8] The government, according to this provision, could not actually increase the number of affordable housing units, even though population constantly increased and with it the numbers of poor. For example, during the three-decade period when Ernie Bohn created approximately 10,600 new dwelling units in Cleveland to replace deteriorated housing, over two hundred thousand African Americans arrived in Cleveland, mainly migrating from the South. Most found themselves crowded into the existing black slums. This single concession to developers—hobbling the construction of new affordable housing—consigned the nation's poor to be constantly displaced as their existing housing was torn down before new housing could be built.

Consider, for example, this blunt 1968 review in which a locally appointed committee assessed Cleveland's public housing effort and cited problems common to all American cities at that time. This report carried with it an implication of racial prejudice that was found in housing programs around the nation. It helped end Ernie Bohn's long and illustrious career.

> Displaced persons constitute one of the groups most in need of public housing. Most inner-city displacement arises from CMHA's [Cleveland Metropolitan Housing Authority] own program of land acquisition, and through the City's urban

> renewal program. Although displaced persons have first priority for new admissions, few displaced persons have, in fact, been accepted in public housing.[9]

This was a national trend, not one limited to Cleveland. Brett Williams's 1992 study of the Anacostia neighborhoods of Washington, D.C., "A River Runs Through Us," chronicles a long history of displacement of poor African Americans for urban renewal projects that included new housing—though never enough to house all the residents displaced by the new projects—and transportation improvements to serve suburban commuters. Although new housing strategies such as Section 8 rent subsidies and state-mandated affordable housing quotas for developers were instigated in recent decades, the problem of displacement continues.

While grudging in its provision of low-cost homes or even its planning oversight of its growing settlements, American culture continued to sanctify speculation. After World War II, in a remarkable triumph for speculation—a habit that a younger U.S. government had fought—NAREB convinced the feds to eliminate "the 'quick-profits' tax that discouraged people from buying a home that they intended to quickly resell for profit."[10]

With the real estate and building industries gaining strength, a new era dawned. Between 1946 and 1953, lobbyists such as William Levitt, whose name is synonymous with tract housing, convinced the government to use FHA funds to back loans for the private construction of more than ten million homes,[11] with the first of the Levittowns featuring more than seventeen thousand homes.

The creation of a speculative housing market was an economic development rooted in several trends. First, speculation itself had come to be seen as a democratic opportunity rather than something that stymied democratic opportunity, as the Founding Fathers or Henry George had understood it. Second, government support for the incipient housing industry was seen as a way to provide both homes and jobs. Third, with the rise of indoor plumbing, electricity, and quarters for the automobile, home construction had become a more complex and specialized enterprise—subject, as agriculture, to economies of scale. Fourth, the U.S. population had reached a size that introduced millions of new buyers to the housing market every year. Finally, as is often pointed out, the growing popularity of the automobile made new areas of cheaper land available for housing. Large tracts of inexpensive land in outlying areas that were previously too remote from urban job centers suddenly fell within commuting distance of those jobs, thanks to the car. This added to the economies that allowed builders such as William Levitt and Ben Weingart to create inexpensive tract housing. And it anointed cheap exurban land with the sanctity of the American Dream.

Levittown, Pennsylvania, opened in 1947. Similar tracts of housing were appearing around the country. The lyrical book *Holy Land*, by D. J. Waldie, details the construction and evolution of Lakewood, a thirty-five-thousand-acre tract of lower-middle-class housing in Southern California begun in 1950, a vast grid of eight houses to an acre. Like many other developers, the creators of Lakewood developed not only housing but schemes by which the federal government would finance the construction of the housing they built, as recounted by Waldie.

> Under Section 213 [of the National Housing Act], the FHA would provide 100-percent financing for construction, but only if the houses were built by a nonprofit cooperative of property owners.
>
> The amount of land one of these cooperatives could develop and the size of its FHA loan were limited by law. The maximum number of houses that could be built under section 213 was 501.
>
> Section 213 was a New Deal program. It assisted rural communities by encouraging property owners to organize a nonprofit building association to put up affordable housing.
>
> Boyar and Weingart used Section 213 to finance the largest suburban development in the nation.
>
> There was nothing illegal about it.

Following these early leads, a veritable flood of tract housing tided the returning veterans and others onto the rungs of the middle-class ladder and shaped the American landscape with which we are all so familiar.

The mass production of housing, along with the home loan guarantees offered to veterans, initially made housing more widely available in the same way that the mass production of smaller goods had made cheap manufactured products such as crockery more widely available in the nineteenth century. Between 1940 and 1960, owner-occupied homes in the United States increased about 18 percent—to 62 percent of all homes. However, over the ensuing forty-five years between 1960 and 2005, the rate of homeownership has been slowly nudged up only another 6 percent, to the current record of 68 percent. (Eight percent of those homes are mobile homes. Ten percent are apartments or townhouses. These numbers have also increased slightly as a percentage of the type of homes people own.) That leaves roughly a third of the nation renting.

Why did the benefits of mass-produced housing taper off after 1960? Is it because all of the land within a reasonable commute of our large job centers has been used up? Is it because an extravagant pattern of development wrote its own doom by taking up too much land and consequently raising home prices on the remaining land out of the reach of average Americans? The answers to these questions foretell the quality of our lives in twenty-first century America.

CHAPTER 11

THE RISING PRICE OF THE AMERICAN DREAM

Knowing about the quality of housing in the United States is essential to understanding the quality of life in this country.

—U.S. Census Bureau, *Population Profile of the United States: 2000*

The price tag on the American Dream is rising every moment. According to a recent report by Harvard's Joint Center for Housing Studies, "Today, nearly one in three American households spends more than 30 percent of income on housing, and more than one in eight spend upwards of 50 percent."[1]

At some point, an individual cannot appropriate any more of his or her income to housing without sacrificing other components of the standard of living such as food, clothing, medical expenses, time, and education. So, in addition to greater financial outlays for housing, the greater burden of housing costs shows up in other, less direct ways. Many of the decisions that people make to save money for housing—such as trying to reduce their bills for food, medical care and insurance, or home maintenance—cannot be easily measured. But some, such as time, can. For example, we work longer hours and we drive farther to our jobs—with commutes stretching in some cases to two hours each way. The newest subdivisions and "communities" being planned around Washington, D.C., and New York City anticipate homeowners who will commute one hundred miles each way. These are the only new homes in these vast metropolitan areas that can be considered affordable to a home buyer of median income.

Another result of high housing costs seems to be the increased personal debt for which Americans are now famous. Edward N. Wolff, an economist at New York University, wrote a paper in 2000 entitled "Why Has Median Wealth Grown So Slowly in the 1990s?" He concludes that, "for the middle class, indebtedness grew sharply since 1983, . . . reflecting mainly rising mortgage debt on homeowner's property."

To reduce housing costs, the less well-off may live more than one to a room. Crowded housing conditions were not uncommon in 1940, but they dropped as more housing was built and homeownership increased. In 1980, however, at around the same time that inequality began rising in America, the number of people living more than one to a room began to nudge slowly upward. The national figure—6 percent—is still quite low. However, this average does not reflect the greater crowding that occurs in denser housing markets—26 percent in Los Angeles, for example, and 24 percent in Passaic, New Jersey.

Why look at denser housing markets? Those markets show the vulnerability of the modern American homeowner, as more and more of us come to live in them. Eighty percent of us live in "urbanized areas"—mainly suburbs—where the majority of jobs are. These places are growing in both area and density—not the density of traditional city centers but the gradually increasing density of inner suburbs, such as Arlington, Virginia, where I grew up, or Cambridge, Massachusetts, across the river from Boston, where I first settled in New England. These closer-in suburbs gradually accommodate more multifamily and condominium housing, more high-rise office and apartment buildings.

The largest and most densely packed metropolitan areas line our coasts, with some major exceptions such as Chicago (with six million people occupying a sort of inland coast around Lake Michigan) and Atlanta. These crowded metropolises are home to the nation's most competitive housing markets, friendly to no income group other than the topmost. They show where we are all heading, and they help to explain the slower rate at which homeownership has climbed in this country during the last forty-five years.

As we might expect, concentrated population causes dramatically lower rates of home ownership. It is in our largest, densest metropolitan areas that homeowners pay the largest percentage of their incomes for housing. (This is also where lower-income people without cars need to live, close to the public transit they rely on to reach jobs.) Median home prices in the ten densest cities and counties listed in the 2000 census average one hundred thousand dollars higher than the national median. About half as many people own their homes in these places compared to those who own them in other towns and counties around the nation—except in the two wealthy counties on the list, with average incomes nearly twice the national average: Nassau County, a suburb of New York, and Orange County, a suburb of San Francisco. Renters pay an even higher percentage of their incomes in these areas.

Our housing burdens correlate roughly with population density and the size of the area over which it spreads, which help to explain the slowdown in rising homeownership in America. These densely populated housing markets act as a kind of forecast of future housing conditions—those that will spread as our population grows, metropolitan areas grow, and population densities increase. Will there still be enough for all? Or will we have to revise this profound cultural assumption?

There are smaller areas—neighborhoods and "metropolitan statistical areas"—that show greater densities and more troubling housing problems, but many of these are slums and are not representative of the experience of the majority of Americans. Other large metropolitan areas, such as Washington, D.C., and Chicago, also have high average home prices, have low homeownership rates (about one-third lower than the national average), and are considered to have housing shortages. But with their relatively vast metropolitan areas, the densities of these two areas do not catapult them into the top ten.

The number of "house poor"—people paying over 28 percent of their income for housing—has also increased nationwide since 1980, about the same time that inequality started its still continuing climb in this country, according to census figures. The government recently responded by raising its recommendation concerning the percentage of income people should spend on housing to 30 percent. When I graduated from college in the late 1970s, people tried to allot a quarter of their income for housing. The census now counts 42 percent of homeowners paying over 24 percent of their income for their homes.

Here in Boston, one of the toughest housing markets in the nation, the repercussions of the rising cost of housing have spread to every border of the state. A 2004 study by MassINC compared Massachusetts housing costs in 1999 and 2003 with the Massachusetts median household income, an income higher than the national median. The study calculated that the most affordable home was two and a half times the median household income. By 2003, "less and less of the state—and virtually none of eastern Massachusetts—was in the price range of middle-income Massachusetts households. In just three years, the number of communities in the most expensive category jumped from 43 to 106." This trend, the study concluded, "raises troubling questions about the future of homeownership for the middle class."[2] But all too quickly, those statistics went out of date. The next year, in 2004, a *Boston Globe* article reported that another thirty-two Boston-area communities had dropped out of the affordable category. This left only twenty-seven of the four hundred or more communities within about a two-hour drive of Boston in which someone making the median income could hope to buy.

At the moment, there is talk that some housing markets have "bubbled" out of control and that housing prices will come down. This indeed occurs periodically, mainly in crowded regions as growth and the speculative response to its demands both heat up. My husband and I bought our house in 1989, just before home prices in Boston dropped by a quarter to a third. But it took only a few years for the 1989 value of the house to return, a few more years for the value to double. A few years after that, our house was worth three times what we paid for it. These vicissitudes are small in the long-term rise of housing costs.

WHO ARE WE REALLY SUBSIDIZING?

In competitive markets, renters pay more for housing than owners, so they are squeezed more tightly. Nationwide, renters pay an average of 29 percent of their incomes for housing. In the ten densest housing markets described previously, renters are the majority, often by two to one. They pay up to half of their income for rent. After paying for housing, Americans in the lowest fifth of the income distribution have only about $250 a month left to spend on food, health care, transportation, and clothing.[3]

Undoubtedly, the third of the population with the lowest incomes suffers most—those whose incomes are no match for the rising costs of modern housing. Some scholars have been willing to point out that the housing crisis is really an income crisis—in a nation that has not raised its minimum wage since 1997, while housing costs in most years, in most markets, rise annually, sometimes steeply. Jane Jacobs, author of *The Death and Life of Great American Cities*, made this point, and Professor Robert Fishman of the University of Michigan has elaborated it.[4] Seen in this light, and in light of the historic struggle to provide decent housing in Cleveland and around the nation, housing subsidies such as Section 8 and public housing are subsidies to employers who cannot or will not pay their employees enough to live on and also are subsidies of our particular style of land development and of the American Dream itself.

To show the gap between housing costs and wages, the National Low Income Housing Coalition (NLIHC) calculates what it calls a "housing wage." The housing wage is "the hourly income necessary (at 40 hours per week, 52 weeks a year) to earn enough income to afford the fair market rent for a 2 bedroom unit," assuming that only 30 percent of income can be spent on housing. In the report, *Out of Reach 2003: America's Housing Wage Climbs*, the NLIHC looked at the wages a person must earn to rent a decent two-bedroom apartment at fair-market value in the housing markets in every state. The current nationwide average housing wage is more than double the minimum wage, but it is much lower than the wage needed to buy a home in any of the crowded, job-rich housing markets of America's largest cities. For example, no county in Northern Virginia that could be considered a suburb of Washington has a housing wage lower than twenty-three dollars an hour. This eliminates not only the poor from these suburbs but the teachers, firemen, and other city workers who work there. And every year, the income needed for the same housing rises.

> In 1999, the national two bedroom housing wage was $11.08; in 2003, the national housing wage is $15.21, a 37% increase. There is nothing on the horizon to cause us to think that rents will not continue to rise. Indeed, the loss of modest rental housing stock continues, as market forces drive up the cost of housing and government fails to intervene to level the playing field.

Regional economic vicissitudes may temporarily change the housing picture for this income group from time to time, but they will not reverse the pervasive trend of the nation. For example, as the NLIHC report relates, housing costs did decrease in nine counties in 2003. But those counties "are home to only 13% of the nation's renter households, while 43% of renter households make their homes in [the 1,371] counties that saw an increase of $20 or more" in rental rates. And, as it has throughout history, the cost of housing continues to rise with increasing concentrations of population.

The rise in income gap since the mid-1970s, exacerbated by an economy driven by technologically sophisticated specialties, contributes to the housing problem. As Fishman notes in analyzing the demise of public housing, public housing was supposed to support most of its enterprise through collected rents, an ever-more unrealistic idea. "Increasingly, public-housing tenants were stranded in the low-wage, intermittent-employment or welfare section of the economy, while construction workers and even the lowest-paid employees of a housing authority enjoyed steady work at rising pay." This meant housing authorities operating at such low levels that monitoring and maintenance of housing fell to lower and lower levels.[5]

During 2005 and 2006, Congress considered, and is still considering, legislation that will establish a national fund for affordable housing, which would derive its income from a tiny percentage of the after-tax profits of Freddie Mac and Fannie Mae. And this would indeed be a triumph for the quality of life for a third of Americans. However, it is important to note that these subsidies support businesses that refuse to pay their employees enough to live on as much as they support housing for our citizens with lower incomes. These programs are small and inadequate efforts to minimally redistribute the resources that our particular brand of economic development has taken from the common pot over time.

CROWDING OURSELVES OUT OF HOUSE AND HOME

As diminishing resources always do, the scarcities of decent dwellings in intensifying urban housing markets affect the poor first. But the housing markets in densely settled places show us how our fate is linked to, and in some way presaged by, the fate of the poor, as the American population continues to grow. We must ask why the burden of housing is growing for all Americans.

Housing starts are now slightly higher than in 1950, when a smaller U.S. population was expanding at a faster rate—just under two million then and roughly two million now, which some analysts consider healthy and some consider scant.[6] Since fewer people now occupy each American house; the same number of homes house fewer people. And these homes are more expensive—that is, as we have just examined at length, they consume a higher share of the American income than ever before. And why is that? Is it because home builders now face the rising cost of land, caused by population growth coupled with decades of extravagant, wasteful development patterns—along with, to a lesser extent, higher labor costs plus expenses related to building in crowded environments? According to Nicholas Retsinas, director of the Joint Center for Housing Studies, the rising cost of land is indeed the main factor in rising home costs. According to his center's 2003 report, developers can no longer afford to build affordable housing. The public, the report concludes, must increasingly subsidize housing.[7] And according to Finley Perry, president of the Home Builders Association of Massachusetts, explaining "barriers to producing single-family homes" in Massachusetts: "There is no incentive for the home-building industry to do anything at the starter-home level. Land is so expensive, you can't really afford to put an inexpensive home on it."[8]

The rise in land costs is a historic trend in America. In this century, housing costs made their biggest leap in comparison to wages during the 1970s, when the value of average income stagnated while housing prices climbed, and they are now doing so again in the first years of the twenty-first century, as they have done to a lesser extent in most years in between. Coincidentally, around 1976 measures of the gap between rich and poor started to rise again. This fact holds implications not only for the poor but for all of us.

Witnesses in the development field—though not all developers—confirm the impression that development is becoming a more expensive, contentious prospect. Henry Cisneros, former secretary of housing and urban development, informally answered questions after a 2003 lecture he gave at Harvard on the housing needs of lower-income

Americans. There, in comments afterward, he remarked that the obstacles developers face in urban areas—facing constraints from previously built-on sites, regulation, from tight public oversight from sometimes warring constituencies—cause them to try to recoup their losses in suburban developments. Cisneros added that large development companies had to consider their shareholders when calculating profits.

A developer in Loudoun County, John Nicholas Jr., also told me that development was becoming more difficult there, as "the tenor of the debate" over development became more shrill and as development increasingly required larger and larger financial commitments, insupportable for all but the "deepest pockets." More difficult development conditions, in other words, favor large corporations, just as changing agricultural conditions have favored large, corporate farms.

When developers look at the difficulties of increased regulation, brownfield redevelopment, and vocal opposition to growth, they sometimes see regulators, conservationists, preservationists, and citizens as the problem. And when citizens and regulators look at the rate and quality of suburban growth in this country, they often fault developers. But manipulating all players on this stage are the forces of population growth and economic strain that cause frustration on both sides. The American Dream is at the intersection of these factors.

The sweetness of the dream and the alacrity with which it has been sold to us for so long with minimal planning oversight have both contributed to our difficulties. Americans have shown only mild enthusiasm for even a slightly more compact version of the dream. The milder expressions of "New Urbanism"—subdivisions with smaller lots, a more tightly woven street grid, and some commercial and community offerings—do not really save enough land to reform our suburban development habits. But even those garner only about 10 percent of the new home market, although these ideas may still be gaining ground. Reforms to our system of development will have to include multiple approaches. Increased shareholder responsibility would help, especially as shareholders in money market–type funds or pension funds are now fairly blind to what they are supporting. Greater regional planning oversight would help, along with firmer standards of land use at the local and regional levels. Tax incentives, along with a national public relations campaign explaining the issues and promoting other housing options, would help. And higher standards of public and private investment in public spaces would help—providing not only adequate recreation areas but shared beauty on the level of, say, Paris, to raise both our standards and appreciation for public space.

When the U.S. government conducted its first census in 1790, there were about four and a half Americans per square mile. There are now eighty of us per square mile. Although this is lower than European densities, and far lower than third world densities, Americans have yet to embrace the greater regulatory controls that help to determine the uses of resources in some European countries.

Many of us demand the American Dream, and many of us sell it. Home building is therefore a very large industry. In addition to increasing competition for land, it has increased our competition for water, with housing development heedlessly flourishing in the driest parts of the West—though new houses there now sport less thirsty landscaping—and draining water from farms. The dream also consumes oil. More than half the oil we use fuels the transportation sector, and supreme in that sector is the automobile we use to navigate our sprawling suburbs.

Perhaps most surprising is the degree to which home building has begun to edge out other uses of land. For example, home building now competes for land with the agricultural, timber, minerals, and oil industries and sometimes even tourism.[9] While we pay a high individual cost for the American Dream in terms of rising home prices, the extent to which housing uses certain resources and reduces our access to others is the high collective price we pay as we house a larger population.

The way we use land is the product of a long history, a natural abundance, and an economic system that rewarded speculation and enterprise. These factors—American history, abundance, and economic catalysts—continually unfolded together at the frontier. And while we have considered parts of the colonial frontier and the Midwestern frontier, the West has proved to be the greatest crucible in which these powerful cultural forces have combined.

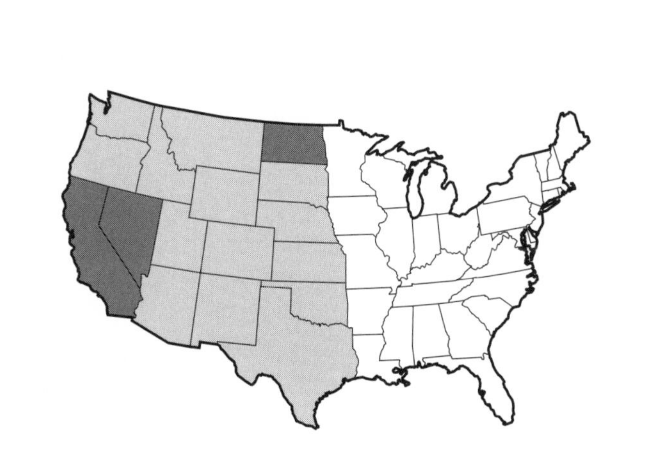

SECTION 4

Land and Enterprise

(THE LATE FRONTIER: THE RACE FOR RESOURCES)

. . . the prairies
Slide out of dark; in the eddy of clean air
The smoke goes up from the high plains of
Wyoming;
The steep sierras arise; the struck foam
Flames at the wind's heel on the far Pacific.
—Archibald MacLeish, "American Letter"

CHAPTER 12

LAND AND LIVELIHOOD IN THE WEST

[T]he frontier is productive of individualism. Complex society is precipitated by the wilderness into a kind of primitive organization based on the family. The tendency is anti-social. It produces antipathy to control, and particularly to any direct control. The tax-gatherer is viewed as a representative of oppression.

The west, at bottom, is a form of society, rather than an area. It is the term applied to the region whose social conditions result from the application of older institutions and ideas to the transforming influences of free land.

—Frederick Jackson Turner, *The Frontier in American History*

We are not all millionaires yet, but hope to be.

—Tasker Oddie, Letter to his mother, Nevada, 1899

The American striking out to gain his or her livelihood shaped the American landscape. The ways that Americans harvested and used resources, and the laws they passed governing these processes, gave us the land we know today. This continual pursuit of prosperity or profit—now codified in our political and economic systems—mainly determines the availability of land for other aspects of our lives—housing, transportation, and recreation—all of which are also industries in their own right.

As Americans swarmed the West, lenient policies encouraged them to take up and use resources. Government policies encouraged mining, grazing, and irrigation of dry land for farms. The federal government continued to underwrite transportation projects of national importance such as the transcontinental railroad. And in creating the legislation that undergirds the American economic system, Congress was guided by the concept that the prosperity of the nation rested on the prosperity of individuals—although we continue to debate the questions of how many individuals, which ones, and under what conditions.

In the mid-nineteenth century, with the opening of the West, American entrepreneurship and the government's relationship to it reached a turning point. Technologies for transportation, resource extraction, and manufacturing advanced dramatically. In service to a large, expanding population, these advances supported strong, interconnected markets. And the annexation of the American West provided a vast field of new resources. National and individual wealth grew. While Astor, Rockefeller, Carnegie, and other tycoons became American icons advertising, via their luxurious existences, the advantages of speculation and monopoly, easterners and foreign immigrants swarmed westward seeking the resources to which they might apply these principles. By the end of the century, the phenomenon of the large corporation was well established, and these corporations were well positioned to influence government.

The western frontier honed and hallowed the privilege of individual access to resources. As Americans settled the West, competition for land intensified, with settlers sometimes sprinting in a chaotic herd to new claims offered by the General Land Office. Speculation and its abuses intensified, including practices such as land-sharking, a staple of Westerns, in which unscrupulous frontier agents sold land warrants to farmers on credit, often at very high interest rates. The farmer had a year to pay. If he fell short, he lost the land.[1] Speculation inspired bold and desperate moves: a man might buy a large area of desert, convinced that he could find funding for large-scale irrigation that would convert it to farmland, if not paradise. These practices, legitimate and otherwise, represented a sharpening of individual economic ambitions that transformed the American landscape. By the end of the nineteenth century, the subdivision, the real estate agent, and the developer were all evolving into their modern forms.

The western frontier differed in other ways from the earliest frontiers of colonial America. By the 1870s, train lines had expanded beyond their early regional webs. They connected western resources to growing eastern markets. They brought people west to swell local economies and markets there. And they did all this with an efficiency undreamed of in colonial days, when fragile ships and hazardous seas connected the Atlantic colonies to England, the Caribbean, and Africa.

The growth of industrial processes and population, the increased national wealth to which those processes catered, and the new proximity of raw materials to markets all had expanded and multiplied commercial energies and the ways in which a clever entrepreneur could take advantage of this growth. The entrepreneur and inventor could create or refine ways to harvest resources, to process, transport, use, or sell resources and products in which they were used. He or she could find new markets or new connections between a product and a market.

This activity accelerated during the late nineteenth and early twentieth centuries as the United States forged a new national identity, and the speculator and entrepreneur and the wealth they generated found newly revered places in that identity. Americans accepted and popularized this race to wealth. They celebrated the stories and the icons that showed the possibility that even individuals from humble backgrounds might take advantage of the system to get rich. The observation that the vast majority of the population would never attain that level of comfort gave rise to labor movements and other leftist political movements. The Socialist Party at its zenith garnered about 6 percent of the popular vote for its presidential candidate in 1912 but thereafter weakened and met increasing hostility as the popular reverence for individualism and wealth was fanned by growing industries of corporate public relations and advertising.

While the early federal government had tried to limit speculation through its land policies, both government and the culture increasingly rewarded speculation and entrepreneurship in the interests of a growing economy. In 1862, the government was handing out free land to settlers through the Homestead Act and still trying to keep the speculator from grabbing agricultural land. However, other federal policies related to grazing, mining, and timbering, while similarly designed to encourage the rapid taking up of land, resulted in more careless use of the land for commercial ends.[2] Subsidies of railroads encouraged speculation in both land and railroad companies. Land grants to railroads gave them a hand in, and an incentive for, promoting development and engaging in land speculation.

Buying large land tracts and selling off lots was almost as easy as finding gold nuggets in a stream. In the middle of nowhere, in places such as Greeley, Colorado, speculators laid out a dirt grid suggesting streets and sold lots to eager newcomers. A continuous stream of veterans' land warrants went west through brokers to fuel speculation, gaining value during the lean years when less public land was put on the market.[3]

Without this honing of American values and American economic activity to reward speculation in the crucible of the West, the modern land developer, whom we shall meet later in this chapter, might be a different animal. The modern American landscape might be a different place. Our daily lives might be routed differently within that different place. Within that different landscape, the housing of our population might be viewed differently. Instead, we find ourselves struggling to reconcile the individualist values of a bountiful, scantily populated frontier with the modern need to conduct and optimize our daily lives under increasingly crowded conditions.

THE STAGES OF THE FRONTIER

Under frontier conditions, resources are more or less available for the taking. Frederick Jackson Turner outlines the stages of the American frontier in terms of resource extraction. He mentions the Atlantic fishing frontier and then the fur-trading frontier, which blazed across the continent. The rancher and miner follow the fur trader. The rancher, Turner points out, can use unimproved pasture and needs only primitive transportation routes since his product can "walk to market." Finally, according to Turner, comes the farming frontier, in which soil is the resource exploited, soil that lies in abundance waiting to be taken, cleared, and tilled.

In all cases, the combination of abundant resources, enterprise, and hard work created the dynamics of the frontier. The early fisherman and explorer found cod so plentiful along the North Atlantic coast of America that they joked about being able to cross the water on the backs of the fish. The fur frontier went on in some northern states for centuries. The earliest prospectors found precious nuggets scattered on the ground and in streams. And rumors of ever more fertile land trickled back east to tempt the farmer westward.

However, in each case, the individual's access to these resources diminished dramatically over time. In the case of the fisherman, overfishing depleted stocks on both coasts. Fur-bearing animals disappeared, and fur farming grew up to replace trapping just as fish farms are a growing trend today. Where loose gold was once found on the ground or just beneath it, companies later mined substantial veins of gold captured in quartz. Eventually the most expensive mining processes were necessary to extract infinitesimal ore deposits among large quantities of rock. Today, gold mining operations dig up huge quantities of rock and treat the crushed rock with cyanide to yield tiny flecks of gold, leaving behind cratered landscapes dotted with mountains of rock and streaming with poisons. Such processes are available only to the heavily capitalized corporation. The history of the extraction of other minerals is similar.

Oil extraction, long a corporate enterprise, is undergoing a similar evolution. For example, dwindling supplies of oil have encouraged the development of the expensive technologies that make offshore exploration and drilling possible and drilling in arctic climates. Next may be the development of technologies that allow oil to be extracted from the sands or shales with which it is often deposited, while the price of oil climbs to support these technologies,[4] leaving the individual to be only a cog in the machinery of resource extraction.

In agriculture, on the other hand, the lack of opportunity for small farmers has more to do with the land's scarcity and cost and the development of food distribution systems in which the individual cannot compete against the interests of the large corporations that have tied up a large share of the market. Seven of the eleven largest private companies in the United States—large enough to be comfortable operating on their own revenues without going public—deal in food. Their combined revenues totaled about $145 billion in 2003.[5]

On the frontier the human traffic created by the search for these resources also created commercial opportunities. This traffic fed the growth of cities such as St. Louis on the Mississippi and Kansas City on the Missouri, and it gave rise to many smaller towns and outposts along rivers, wagon train lines, and finally rail lines across the West.

This continual migration eventually settled even the less desirable lands. In the case of Nevada, for example, European American foot and horse traffic crossed the largely barren landscape for roughly a century before the California gold rush swelled the stream of migrants to a size that could support a trading post. First, the Spanish had explored Nevada for routes to California in the late eighteenth century. Furriers crossed the state and took what little they could from approximately the 1820s to the 1850s. In the 1840s, immigrants to California began to cross Nevada from Utah. By 1850, with the gold rush under way, mineral discoveries occurred in Nevada. There was finally enough traffic at this point to inspire the establishment of a trading post at Genoa, near Lake Tahoe and near the first mineral discoveries in the future territory.

THE MINING FRONTIER

A humbling grandeur awaited immigrants to the West. Daunting expanses of desert, mountain, and prairie met the pioneers, and climatic extremes dogged them. Much of the West was too dry to farm or would become so in the next climatic cycle—that is, the current one.[6] Wildlife abounded in the north of this area, timber in the far northwest. The plains and grasslands supported cattle. And beneath the forbidding landscape of desert, mountain, dry hills, and prairie lay the vast mineral wealth of which earlier English monarchs had dreamed—not only gold and silver but copper, lead, zinc, mercury, and an array of semiprecious stones that the native people had mined for weapons and jewelry.

After the United States acquired its western territories, mineral discoveries shaped the western map. The discovery of gold in California in 1848 led not only to the immediate and exponential growth of California—from a non-Indian population of about fourteen thousand in January 1848 to nearly one hundred thousand by the end of next year—but to the discovery of gold in Nevada by Mormon forty-niners en route to California. The Mormons' early discovery of gold led within ten years to the discovery of the Comstock lode, a Nevada silver mine so rich that it drew national attention and made many investors rich.

Gold was discovered in Colorado in 1859, drawing settlers there.

The discovery of gold in Montana in 1862 and in Idaho in 1863 initiated settlement in those states. Also in 1863, the first formal claims for gold and silver mines were filed in Utah by non-Mormon explorers, despite Brigham Young's efforts to discourage mineral exploration to avoid bringing outsiders to the state. In 1874, the discovery of gold in the sacred Black Hills of the Dakota tribes led to settlement there and helped complete the demise of the Indians. And under the dry wastes of Nevada lay so much ore that Nevada today is the number one producer of both gold and silver in the United States, third largest producer of gold in the world—and, not coincidentally, the gambling capital of the nation.

As mineral discoveries continued around the West, Congress saw a way to encourage settlement of the vast western territories while promoting the national economy. The 1872 Mining Law stated that any American citizen, or anyone declaring his or her intention to become an American citizen, could explore the public lands for minerals, could take what they found, and could purchase the land, surveyed or not, for five dollars an acre. Congress had not only found a way to encourage speculation and shape much of the West in the speculator's image. It had created what would eventually become a huge giveaway to large corporations whose mining processes would devastate the environments in which they operated. Each successive process of mineral extraction involved greater environmental destruction. For example, the cyanide extraction process uses up to eighty tons of rock to extract an ounce of gold and creates pits up to a mile in diameter.[7] And each successive process involved larger corporate structures to underwrite the increasingly expensive extraction methods.

But we are getting ahead of our story. In our story, the mining frontier further fueled the speculative fever of American economic growth and enshrined the concepts of luck and gambling in American mythology—presaging their role in shaping our modern landscape. That story lies in the Old West.

Forty-one years after the gold rush, in 1890, San Francisco was the largest city west of the Mississippi with 250,000 people. Men, women, and families were still heading west looking for luck or land. The Homestead Act made land available to single women, considerably advancing American women's property rights, although those rights vanished with marriage. And clever young men came west for the opportunity to sharpen their wits making western resources into a livelihood or, if possible, a killing.

TASKER ODDIE GOES TO NEVADA

Tasker Oddie, born in Brooklyn in 1870, fell under the spell of the West as a sickly adolescent. Raised in nearby East Orange, New Jersey, where he attended the public schools, he was sent west for his health. His family's connections enabled him to help out on a Nebraska ranch for a couple of years to regain his strength. The cure worked and left Oddie with a lifelong passion for the western outdoor life.

After he returned east, Oddie took a job in business in New York and attended night classes in the Law Department of New York University, graduating and gaining admission to the bar in 1895. At the age of twenty-seven, the hard-working, energetic, and prematurely balding young man found an opportunity to travel west as the secretary of the Nevada Company, which had mining and railroad interests in central Nevada as well as a supply store, allowing company executives to dabble in several related aspects of a development trend, much as horizontally structured corporations do today, though the company was not highly successful.

Before Tasker Oddie arrived there in 1898, the discovery of gold and silver in Nevada had fueled fifty years of growth. Nevada had earned territorial status by 1860 and statehood in 1864. Mines had multiplied during this time, though Nevada's population remained sparse for decades, with a good sampling of ranchers, ranch hands, prospectors, miners, mine administrators, a few lawyers, and those connected with the operation of trains, banks, or stores and supply lines. The meager offices of government offered a few posts, and the occasional church called for a rector.

In 1860, only seven thousand people were scattered over Nevada's 109,000 square miles. The next year, 1861, sixteen thousand people occupied Nevada, still sharing the territory with Indians and emigrants passing through. Ironically, that same year, "games of chance" were outlawed in the first Nevada Gambling Act. In announcing the statute, Governor Nye took the opportunity to revile gambling as "the worst" "of all the seductive vices extant."[8] Other antigambling statutes followed in later years, but, as the life of Tasker Oddie shows, gambling was not just a game but a way of life in Nevada and much of the West. A man like Tasker Oddie gambled his youth, his energy, and his earnings on achieving independence and riches. It is hard to say whether independence or wealth mattered more to him. And Oddie was surrounded by people whose lives reflected his own search. In letters home, Oddie himself mused on the lack of cultural life, the gambling at Indian fandangos and how little other than work his fellow settlers had to occupy themselves.[9]

Oddie arrived in Nevada as a lawyer-accountant and clerk and made a good impression, even uncovering a bit of fraud for the company. Within his first year, however, he made the transition, with his employer's blessing, to shopping for his own mine. He confessed, in his letters, to "ambition." Along with the fond memories he carried of his outdoor life in Nebraska, Oddie seems to have contracted the widespread speculative fever of the day. Even though his mother and sisters in New Jersey were partly dependent on the money he sent them, the otherwise thoughtful young man, who wrote home constantly, withdrew this support to purchase his first mine and suggested they borrow from his aunt "till I can make something out here."[10]

Oddie arrived in the West when the most accessible resources had been taken up. Ranchers with thousands of head of cattle had become so thick in much of the West that the herds had decimated the public lands on which they were allowed to graze. To keep out competing herds, ranchers had begun to use the recent invention of barbed wire to cheaply fence in tens of thousands of acres of public land for their own use, as Oddie described in his letters. Many mines had already been opened in Nevada and the purest ores removed. Such were the first mines in which Oddie bought interests. Regarding his first purchase of a mine—a quicksilver or mercury mine—he wrote to his mother, "The old man who owned it took about 100 lbs. of the best ore from the surface and worked it in a crude retort, and got 15 lbs. of pure quicksilver from it. That is higher than the average will be but it shows that the ore is good."[11]

From the point of his first mine purchase, Oddie's letters meticulously detail the hardships of his new life—of living outdoors and trying to develop a camp at his new mine, which was located at an elevation of 11,500 feet and to which there were no roads. Because Oddie arrived late on the scene, after the best had been taken, his labors were endless and not entirely fruitful. He seemed to spend most of his time chasing or replacing lost or lamed packhorses and mules in scenes that read like slapstick comedy. "To come back to the mules," he wrote, "more perverse, stubborn, mean devils never lived. Their sole aim in life was to torment the soul out of me every day."[12] Except for the mules, his complaints were mostly good natured, concerning such matters as having to eat his own cooking with only a campfire to work on. He seemed to love his "healthy" outdoor life.

In addition to chasing horses around his mountain, Oddie's other preoccupation was calculating the riches he might make if things went well, which they never quite did at the first mine. He bought a few other mining interests or options, but nothing came of them until September 1900. His mother and sisters had gone through "Auntie's" loan and were ready to take in boarders to try to keep their house when finally a prospector offered him an interest in a fresh find—the discovery of "a group of ledges that promise to be very rich."[13] Those ledges became the Tonapah silver mines.

At about the same time, Oddie, at the age of thirty, became district attorney for Nye County, beginning a modest political career that was to give him a small but steady income when his mining ventures failed to profit. But for a time, Oddie did in fact become rich from the Tonapah mines, after further hard work to develop the mines with his partners. He brought his mother and sisters west, and his mother remained with him until her death. He gave lavish parties; lived some of the fantasies that he had held close during his mule-chasing days; and eventually married.

But when the money from Tonapah was gone, Oddie never found another mine. Time had passed, and resources had been taken up. Likable, smart, and well educated in comparison to his contemporaries, Oddie served in the state legislature, served as governor from 1911 to 1915, and was finally elected to Congress for two terms, but he continued to be a gambler at heart, as a letter to a political supporter in 1920 shows. Defending himself as his debts were becoming a political issue, Oddie argued: "I have developed large numbers of mines and prospects in this state and have spent large fortunes on them and have been constantly investigating new prospects for years past and intend to have another good one before I get through."[14] This sounds like the twenty-nine-year-old Oddie of years before, writing his mother of his first mine: "If the mine turns out as I think it may, it may make a millionaire of me yet,"[15] and later, "There is a chance that I will be able to pay off the entire mortgage before a great while."[16]

Men like Oddie abounded in the West. Their tolerance for risk varied. Their health and capacity for hard work varied. Their luck in finding unclaimed resources varied. In different regions and at different times, their stories changed with the resources and markets they found, but such men, and occasionally women, are a feature of frontier history, even through the rise of the recent development frontier where builders rush to construct housing or shopping centers on cheaper land in the American suburbs.

By the end of Tasker Oddie's service in the Senate in 1933, another clever young man named Harold Pollman was growing up around the prairies of North Dakota, helping his father, Paul, to harvest the last gleanings of the long Dakota fur frontier and preparing for the frontiers to come. Young Harold was learning the ropes of his father's business and, more important, learning the lessons of resource harvesting and markets. Harold would be in on the North Dakota oil discoveries before he even finished law school. And when the home loans authorized for returning GIs began to kick in, Harold moved to Texas, ready to build homes for veterans and their families.

The stories of the Pollmans in North Dakota and Tasker Oddie in Nevada describe different individuals in different settings, but there are similarities. Theirs is not the story of the early frontiers—of immigrants going west to stake a homestead claim or to find cheap land, free grazing range, or loose gold. Between them, Tasker Oddie and the Pollmans arrived late on an assortment of frontiers—minerals, fur, and oil. But through native cleverness and, in the case of Oddie and Harold Pollman, law degrees, they found themselves well suited to take advantage of what resources remained. By the twentieth century, diminishing resources had already begun to sort the more able from the less able more discriminately and the educated from those who were not, as Harold's story shows.

CHAPTER 13

THE DAKOTAS

The object of your mission is to explore the Missouri river, & such principal stream of it, as, by it's [sic] *course & communication with the waters of the Pacific Ocean, may offer the most direct & practicable water communication across this continent, for the purposes of commerce.*

—Thomas Jefferson, instructions to Merriweather Lewis, 1803

This little fleet altho' not quite so respectable as those of Columbus or Capt. Cook, were still viewed by us with as much pleasure as those deservedly famed adventurers ever beheld theirs; and I dare say with quite as much anxiety for their safety and preservation.

—Merriweather Lewis, Fort Mandan, North Dakota, 1805, beginning the westward exploration

The fur frontier endured for over two hundred years in the Dakotas. It was already old when President Jefferson, echoing George Washington's earlier concern with "water communication" to serve "commerce," instructed Captain Merriweather Lewis to follow the Missouri River. It was older still when one of Custer's soldiers found gold in the sacred Black Hills of the Sioux in the 1870s and newcomers stampeded the Dakota Territory. And the territory still had furs to offer, though many fewer, by the time farmers had claimed the best lands and celebrated statehood.

The area of North Dakota where Harold Pollman grew up was one of the last agricultural frontiers in the union for a number of reasons, including the brutal winters that produced such luxurious pelts on the prairie mammals. In addition, the Louisiana Purchase that brought most of the northern plains into U.S. possession had not included the divinely fertile land around the Red River that separates the Dakotas from Minnesota. The British retained claim to it until 1818 as part of the watershed of Hudson Bay in Canada.

British fur traders of the Hudson's Bay Company, plus French and Canadians, had been working this northerly area of the Dakotas—and beyond—since the early seventeenth century. They favored beaver, fisher, otter, marten, and mink, as well as the bear, wolf, and fox. And they favored the independence of a peripatetic life. Many built camps or trading posts. Many took Indian wives. By the nineteenth century, their numerous descendants filled out the Métis Nation—a mixed European-Indian population around the Red River.

In 1804, Lewis and Clark paddled up the Missouri. They passed through the heart of what would become South Dakota and built Fort Mandan on the bank of the Missouri, near present-day Williston, North Dakota, in the southwest quadrant of the state. There they found their invaluable translator and ambassador, Sakajawea, and her French trapper husband. They waited out the brutal winter in the fort—recording their daily, almost familial interactions with the docile Mandan Indians and recording temperatures plunging as low as forty-four degrees below zero—before heading west in the spring.

The prairie that greeted these adventurers was a vast, soft-hued ocean of waving grasses. The hypnotic eternity of the plains was not just expansive and disorienting but profoundly fecund. The soils of these plains, enriched by "the alluvial deposit from primeval lakes," as Frederick Jackson Turner wrote, "surpass in fertility any other region of America or Europe except some territory about the Black Sea [from which future immigrants would come]. It is a land marked out as the granary of a nation."[1]

But for centuries before Americans could transform it into their granary, the soil supported native grasses that provided both food and habitat for myriad animal species. Abundant shallow lakes provided perfect watering holes.

While the country was settled and the Revolution raged, the fur-bearing animals of the future Dakotas enriched the French and the British ranging out from Canada in the seventeenth and eighteenth centuries. By the early nineteenth century, the Americans had joined the competition. German immigrant John Jacob Astor had by then included the Dakotas in his near monopoly of the American fur industry through his American Fur Company.

In 1834, having skimmed the cream, Astor sold his fur business. He invested his profits in Manhattan farmland and watched them turn to gold as his land became the heart of New York City. The fur trade in North Dakota continued without him, and, for a while, plenty of pelts remained. In 1851, a fur trader in North Dakota was able to ship to market the furs of four thousand lynx, eight thousand martens, and one thousand minks, with a dozen silver fox furs thrown in.[2]

Though the North Dakota prairie provided a superb habitat for ninety species of fur-bearing animals, the majestic plains were first and foremost a vast pasture for a more majestic creature, the buffalo. So plentiful were the buffalo of Dakota Territory that a man could ride through them for days on end or gaze over the endless prairie and see nothing but buffalo, as did Lewis and Clark, and later, Charles Cavileer, the first customs inspector in the northern county of Pembin, in 1851, from atop Turtle Mountain. "I could see for miles and miles," he wrote, "and the prairie was black with buffalo."[3] Today, all that remains of these majestic beasts is a few protected herds; a small, exotic niche in modern agriculture; and the vast prairie pasture they shaped for centuries through their grazing.

The great buffalo herds, in turn, supported several cultures of native peoples, who made everything from cooking equipment and clothes to homes from the hide, sinew, bones, and hooves of the buffalo, on whose meat they lived. Change came slowly to the windswept prairies—with their buffalo, their native people, and their harsh winters and hot summers—but inevitably it came.

READYING DAKOTA TERRITORY FOR THE FARMER

The Indians had supplied many of the furs taken by traders over the centuries and in relative peace. The Mandan had treated Lewis and Clark as elder brothers. But the accelerating American quest for land and minerals ended the peaceful nature of the symbiotic commerce of the fur industry. In the 1830s, smallpox decimated the native population of Dakota Territory, but there remained enough Indians to fight for their way of life and the land and resources that supported it. Furthermore, the tribes near Canada enjoyed some British support through the early nineteenth century. The U.S. government formally organized the Dakota Territory in 1861. As Minnesota Indians were displaced and pushed into the Dakotas by white settlement, conflicts intensified.

In the 1850s and 1860s, U.S. cavalry regiments marched into the territory. Shortly before Congress organized the Dakota Territory in 1861, the U.S. Army began to buy trading forts and to build new forts there. Shortly after the territory was established, Minnesota cavalry regiments arrived in Dakota Territory to pursue rebellious Indian leaders who had fought in Minnesota.

The legendary battles of the West were fought in the Dakotas—both Wounded Knee and Little Big Horn—and both sides staged massacres there. The horses that lent grace to the image of the Plains Indians and efficiency to their buffalo hunt as well as the many guns that had been traded for furs lent a fleeting success to the native resistance to the white man's incursions. But they were eventually outmatched by the howitzers and then the Gatling guns that arrived with the cavalry in the 1860s.

In June 1863, when the Seventh Minnesota rode into northern Dakota Territory, Colonel William R. Marshall kept the regimental journal. By July the soldiers sited numbers of the Sioux monitoring their progress almost daily—a growing presence. On July 29, Marshall wrote: "The Indians were in plain view in immense numbers on our west side—part of their train ascending the bluff. These on the bluff were constantly signaling to those at the river by flashing the sun's reflection from a mirror."[4]

The terrors of entering hostile Indian territory, even with howitzers, almost explain the U.S. government's decision to start killing buffalo—which could not fight back—in order to starve the Plains Indians into submission. This policy, undertaken in the 1870s through a bounty system, was highly successful. More western legends were born. Buffalo Bill Cody killed thousands of buffalo and built a career on this feat. And the Indians were left with their way of life as dead as the buffalo.

By this time, good arable land was becoming less available, even on the frontier. The U.S. Land Office had been offering swamplands for a couple of decades and was formulating policies for the arid lands of the West. Though the agricultural sector of the national economy was changing and giving ground increasingly to manufacturing, a number of factors urged the cultivation of the Dakotas. The exceptionally fertile soils, flat topography, and absence of rocks and trees all invited the use of farm machinery and the development of large-scale agribusiness. After Cyrus McCormick had brought his new reaper to Chicago in 1847 to serve the midwestern grain fields, further mechanization of agricultural processes followed, for which the Dakotas' large, flat fields were ideal.

The soils of the Red River Valley were particularly blessed. Rumors to this effect drew investors who charted steamboat lines up the Red River and built railroads west from St. Paul, anticipating an agricultural boom there. The perfectly timed Homestead Act of 1862—giving 160 acres of land to men and single women free of all costs except a minimal filing fee—invited Americans and immigrants to try their luck at farming on the nation's remaining unclaimed land. And if that were not enough to draw population to the new states, then events of 1873 and 1874 sealed the deal.

THE BONANZA FARM AND THE TWENTIETH CENTURY

The Panic of 1873 helped to bankrupt the Northern Pacific Railway, which had received from the U.S. government one of the largest land grants in history. Upon the company's failure, a large quantity of company land became available at very low prices to the railway's investors. Thinking as businessmen rather than farmers, these investors "secured large blocks of land, professional management, and large-scale machinery to create the bonanza farms of the 1870s, 1880s, and 1890s."[5] The planting of the Bonanza farms was perfectly synchronized with other historical developments, including the introduction of mechanical reapers and other farm machinery, the expansion of railroads, and improvements in wheat milling processes. At first, ranchers competed with farmers for land in the buffalo's old pastures, but the drought of 1886 ruined many and made way for more newcomers seeking land, who tended to be farmers.

At roughly the same time that the fertility of the Red River Valley began to draw farmers and investors, a more dramatic discovery occurred in southern Dakota Territory. General Custer led a thousand men into the Black Hills of South Dakota, land sacred to the warlike Sioux. One of Custer's men found gold along a Creek there. As if the gold nuggets had been the keystone in a dam, their removal unleashed a flood. A tide of prospectors and miners swept into the territory, but this onslaught also stimulated rumors of good soils that drew more farmers too. A hundred thousand people immigrated to the territory between 1879 and 1886, with many homesteaders taking up land in the northern part of the territory under the Homestead Act.

Newcomers to the Dakota Territory found a generous land in an unforgiving climate. Annual North Dakota rainfall is just shy of the minimum desirable for farming—twenty inches compared to the preferred twenty-two. Brutal winters required homesteading families to spend much of their earnings on fuel to tide them through the long months of subzero weather. Summer temperatures often exceeded one hundred degrees. In addition to merciless weather, it turned out that a state full of grass could support amazing and destructive plagues of grasshoppers.

The Bonanza farms complicated life for the small farmer and the average citizen. While making tremendous profits for their owners, the Bonanza farms were hard for the smaller farmer to compete with. The Bonanza farm, because of its size, had greater leverage with railroads, manufacturers, and politicians. In addition, the large plantations directed the state's economic growth in ways that benefited the class of agribusiness owners, much as large southern plantations had done. In the colonial South, the plantation owners' land and wealth supplied the English market for cotton, tobacco, and other high-profit crops. The wealthy landowners also bought their luxuries from England. This initially stymied the growth of local economies—local manufacturing for local markets of the kind then seen in the Mid-Atlantic region.[6] In North Dakota, Bonanza farm owners often lived out of state or even abroad. Their product was aimed at the national market. The farms were highly mechanized, limiting their need to hire local labor. Their contribution to population growth and economic growth was therefore limited. And their interests continue to dominate the state political scene. According to a 2004 Republican candidate for state senate from the district including the city of Grand

Forks, "The big farm owners contribute heavily to the campaigns of our candidates for Congress. They pretty much call the shots around here."

Nonetheless, free land and abundant resources made the Dakotas a natural draw for new immigrants for a few decades in the late nineteenth and early twentieth centuries. Norwegians and Russians had already established a foothold in the state toward the end of the nineteenth century, and they came in greater numbers starting around 1905. By 1910, over a quarter of North Dakota's population was foreign born.[7] Available land made North Dakota a far better destination for the immigrants from an agricultural background than did the factories of Cleveland. By 1915, with nearly six hundred thousand residents in North Dakota, roughly four out of every five residents were either foreign born or the children of immigrants.

Around that time, in 1910, a young Ukrainian named Paul Polman (later Pollman) arrived in New York from Russia. Like John Jacob Astor before him and millions of other less successful immigrants, he found menial work in New York City before heading west. There he met an attractive and spirited young typist-bookkeeper named Sallie, whose brother was making a good living as a furrier in North Dakota, selling the coats he made from his store in Minot.

Paul Pollman had learned a little about furs in Russia from his father. This fact, combined with his new connection through Sallie and perhaps his own enterprise and hopes of leaving behind his job as a bricklayer, gave Paul the impetus for a new undertaking. Before long, like Astor, Tasker Oddie, and millions of other anonymous immigrants before him, Paul Pollman was heading west.

The nation spreading out beyond New York when Paul Pollman arrived was not the same nation that had greeted Astor 150 years before or even the one that had greeted Tasker Oddie 20 years earlier. In the two decades since Oddie had gone west, development had proceeded apace. The remaining pockets of frontier had shifted and dwindled. The resources remaining took more than luck, cleverness, and capital to locate and harvest. Pollman would show a canny knack for harvesting the remaining fur species in North Dakota and provided well for his family, but he did not become wealthy. He was able to see to it, however, that his children received better opportunities. Like many immigrants, Pollman combined hard work with shrewdness to build a life from which he could propel his children to a level of prosperity he could only imagine.

Paul Pollman's process of settling in North Dakota was interrupted briefly when he volunteered for army service in World War I. But by 1920, the U.S. Census found him living as a lodger with the Garber family in Grand Forks, North Dakota. By 1930, the census listed Paul Pollman as a hide buyer living in Minot with his wife, Sallie, and two young children, Harold and Laura. He operated a fur business, shipping furs to auction and supplying some of his brother-in-law's furs. Later in the 1930s and for most of Harold's youth, they lived in Grand Forks.

Harold Pollman grew up with the rituals of fur procurement, at least one of which was his father's invention. Paul Pollman cast his eye on the cream stations around the state where farmers brought their milk for sale. Pollman decided to leave a checkbook with the milk buyer at each creamery so the buyer could purchase for him the pelts of the nuisance animals that farmers might have shot on their land. Hares and rabbits, for example—conspicuously absent from the lists of earlier Dakota fur traders—remained to frustrate the farmer, along with the traditionally coveted mink, ermine, fox, and beaver. (Bear had been so thoroughly exterminated that Pollman never heard of one in North Dakota.) Eventually, Pollman was gathering furs from forty cream stations around the state, improvising a rough replica of the frontier trading post network.

Traditional fur trading remained as well, though in a form changed by the degraded status of the Native Americans, now living on reservations. Harold remembers going with his father to the Fort Totten Indian Reservation at Devil's Lake (a perversion of the beautiful Indian name, Spirit Lake) to trade for furs with the Indians as whites had been doing for a good 250 years. Father and son would stay overnight in the mud-brick and thatch-roof hut of an Indian family before leaving again the next morning.

By the age of eleven or twelve, Harold was driving trucks full of furs on the lonely gravel roads of North Dakota up to eleven hours a day, earning the same wage as his father's other employees. The responsibilities of judging and grading furs came early, too, and Harold helped with that task before the furs were shipped east. Life involved hard work alongside his father but also useful lessons in the harvesting of resources and the effort of earning an independent livelihood, lessons that later made Harold a very successful man.

Even Harold's mother used her wits to improve the family's situation. After the family moved back to Grand Forks during the Depression, she fell in love with a beautiful house on one of the finest streets there. She suggested to Paul that she could use one of the bedrooms for roomers, whose rents would cover the mortgage. And so they moved, and two young women who taught school became part of the household.

Earlier, Harold's father had also signed up for land under the Homestead Act. Though they gave up the land eventually, Harold remembers the common procedure that concerned his father and so many of their neighbors trying to own land. "You could stake your claim on 160 acres," he recalls, "and the government took steps to make sure you weren't a speculator. First, you had to go out there and break the sod. Or you could hire someone to do it for you. People who did that were called sodbusters. Then you had to build a house. If you cut blocks of sod, it was something like adobe. You had to build the house and plow ten acres the first year. The second year you had to plow some more, plant a crop, and plant two or three trees, a certain number. The county agent would come and make sure you were following the steps. That was called 'proving up' your claim."

Meanwhile, Paul Pollman's understanding of his fur and hide trade and related opportunities was expanding. He began buying wool from sheep ranchers around the state and became a large supplier of wool to buyers in the Boston woolen mills, shipping as many as ten to fifteen cars of fleece each summer. During World War II, Pollman supplied hides to St. Louis military boot manufacturers working under government contract. Many Massachusetts woolen mills also received government contracts to supply fabric for uniforms, and Pollman continued to supply them with wool.

During all this time, Harold thrived working alongside his father. He considers his early responsibilities transformative, and one occasion in particular left its mark. One day, when Harold was in his mid-teens, the government hide inspector for the St. Louis tannery fell ill. The hides had been brine cured, drained, and prepared for shipment. They had to be expeditiously graded and shipped. The company, trusting Harold to ignore his father's business interests, instructed Harold to grade the hides to government standards and make the shipment to the St. Louis tannery.

Harold felt the honor of this responsibility; the tannery representative believed in both his honesty and his judgment. The incident called forth from Harold his very best effort. It endowed Harold with a tremendous pride in his honesty that lasted throughout his career. "I have felt that burden of responsibility all my life—that burden delegated to me at that time. It stayed with me throughout my professional career. I think that's why I've only had two lawsuits in a fifty year career as a developer." (By comparison, some large developers are constantly plagued by lawsuits. Harold's were both small—involving one dispute over a property line and one over a parking space—and neither went to court.)

Before Harold became a developer, however, and a leader in his field serving on the boards and in the offices of numerous professional organizations, still more real-world schooling awaited him, in the form of World War II. Harold left college to join the air force, serving on fifty combat missions, including one in which his plane was shot down behind enemy lines and the crew rescued by Yugoslav Partisans and smuggled to rescue vessels on the Adriatic Sea.

Harold arrived at the University of North Dakota to finish his college experience no longer a teenager but a decorated veteran. He was now back in Grand Forks with his family, and after earning his baccalaureate, he started at the University of North Dakota Law School. He was still in his first year there when his sister's husband, an oil executive, called with a tip. Harold, his brother-in-law told him, would be extremely well positioned if he could earn his law degree in oil and gas law. Harold hopped on a bus to spend Thanksgiving with his sister and brother-in-law and learn more.

Harold's brother-in-law had alerted him to the early stages of exploration for oil in North Dakota. Energy consumption was accelerating at a formidable rate in the United States. The growing popularity of the automobile in America, coupled with growing industry and expanding use of electricity, was driving exploration and expansion among oil companies. When, after Thanksgiving, Harold returned to Grand Forks, he undertook to transfer to the University of Texas Law School at Austin. There he concentrated on real estate and oil and gas law, a legal specialty unavailable at his old school. He rushed toward his degree as fast as his courses permitted.

Texas was oil country. While there, Harold met the men who were pioneering the western oil frontier, the men who would help to launch his career. Oilman and wildcatter M. B. Rudman was acquiring leases for drilling on thousands of acres of North Dakota farmland, for which he needed legal representatation. And Rudman introduced Harold to Thomas Leach and A. M. Fruh, two indepenedent oilmen in Bismarck, North Dakota, who similarly had acquired leases to thousands of acres of North Dakota farmland above a geological phenomenon that would soon become known as the Williston Oil Basin. With oil and gas exploration converging on North Dakota, Harold graduated just in time to become the first oil and gas attorney in North Dakota. He set up his practice in Williston just in time to become the attorney of record when the first independent drilling rig in the state struck oil in April 1951—at the Williston Oil Basin, the second-largest oil-producing province in the United States. The legislature of North Dakota then hired Harold to codify North Dakota's first oil and gas laws. And in the oil rush that followed, Harold became wealthy.

By May 20, 1951—only forty-five days later—thirty million acres of North Dakota land, more than half the state—had been leased or purchased by oil companies. Harold oversaw the legal work behind many of these sales and leases. Like the early surveyors, he managed to buy some choice parcels himself—for resale to oil companies rather than to settlers—but the process was similar. Not only was Harold becoming wealthy, he was still a young man, and he already had the next frontier in his sights: the development frontier.

THE DEVELOPMENT FRONTIER

The development frontier comprises the smallest pockets into which speculative activity on cheap land—the mark of the later American frontier—can disperse: the suburbs and exurbs around major American cities. As seen in earlier chapters, the development frontier had begun as a spate of land subdivisions in the suburbs of major American cities. On maps of leading American cities from the Civil War to the end of the nineteenth century, the names of stately suburbs such as Oak Park mingle with lists of raw land claimed for subdivision, waiting to be turned to profit: "Canal Trustees Subdivision," "Ogden's etc. Subdivision," "Flourney's Re-Subdivision of Jones & Patrick's Addition," and

"Poyntz's Subdivision," to name a few of the scores listed on an 1860s map of Chicago.[8] Subdividers gradually discovered that the design and construction of homes and the styling of the subdivison as a community helped to sell the subdivision.[9] After World War II, of course, that trend progressed to an avalanche of home building and early shopping center construction. Today, the American real estate and construction industries together generate over $1.5 trillion.[10]

With new wealth, and continuing the easy relationship with resources that had characterized both his father's and his own career, Harold Pollman turned more and more to the real estate part of his practice and to Texas, which, unlike North Dakota with its relatively stagnant population size, gained a couple of million people between 1950 and 1960. Within a short time, Harold Pollman moved to Texas, where he began buying land and building subdivisions.

Harold's understanding of human nature gave him an advantage in the public relations aspect of development and in making the sociological observations that were an important part of the development business. The idea of creating homes for people, a little more prominent when Harold began in the business, appealed to him. He created different kinds of subdivisions to appeal to different clienteles. "I noticed there were a lot of people moving to Texas from New England and they were looking for homes with fireplaces—something you hardly ever need in Texas. So I built Hearthside Homes for those people."

Although Harold's politics run toward the conservative, his social impulses—with long prairie roots—matured over the years into a philanthropic bent, common among his wealthy peers in Dallas. Even in his seventies, he has had the energy and public mindedness to serve on many boards and advisory committees related to home building.

Harold's social concerns appear to give him patience with the many frustrations that face developers in the tighter—and more tightly regulated—land markets of the modern day and keep his insights into his field current. Harold has had a long, satisfying career beginning at the very peak of the home-building and development boom of the early 1950s—the peak of the development frontier.

Though walking with two canes at nearly eighty years of age, Harold recently visited Harvard's Real Estate Development Club to discuss a Dallas shopping center he built some years ago. When a new highway siphoned business and high-end buyers away from his shopping center not long after it was built, it became a blue-collar shopping center. The surrounding neighborhoods gradually became Hispanic. Now Harold is redeveloping his mall for the new population, a process he appears to enjoy. (Harold's wife and daughters are bilingual, and one of his daughters is a consul to Spain.)

Harold is supervising the redevelopment of the shopping center from a large comfortable recreational vehicle that he shares with a personal assistant. From his RV, Harold can remain close to the project. "The fact that Harold oversees every aspect of his developments himself is one of the secrets of his success," observes Professor Rick Peiser, who invites Harold to lecture his Harvard real estate classes twice a year. Harold's involvement is not dutiful but full of relish. He enjoys telling the story of how, one morning, dozens of Hispanic women lined up outside his trailer for mammograms, because of confusion over where the mobile lab from the local hospital had parked.

Harold's enjoyment—heightened by the philanthropic impulses of maturity and the satisfactions of a long, successful career—is not widely shared by developers. Many real estate developers regard the increasing regulations and limited land supply that dog their endeavors with greater animus. And it is not hard to see why.

ANOTHER FRONTIER WANES

Growing population creates economic pressures (rising land prices and cost of living), political pressures (increasing regulation and litigation), and social pressures (raging NIMBYism) that constrain the developer's business as surely as dwindling fish stocks have changed the fisherman's. John Nicholas, a Loudoun County developer with a cattle farm in the western part of the county, laments both the rising cost of doing business and the current climate of "suspicion" that surrounds development. He admits that, while, within his memory, small constituencies have often opposed a given development project, the rise of environmental groups ready to mobilize around these local battles has changed the playing field. Environmental groups, of course, are enjoying popular and monetary support from much of the citizenry in many growing communities such as Loudoun.

In a 2004 phone conversation, Nicholas estimated that "Now it takes four or five years from the time you buy a property until you can begin to build on it." Holding onto expensive land without being able to generate any return on the investment for four or five years is, of course, a task for large corporations rather than individuals or smaller business entities, a sign that the development frontier is nearing its endgame. In real estate development, as in other industries, wealth and connections become critical to the exploitation of resources as those resources recede. In as far as large corporations possess both, to that extent they inherit the resources of the nation. By the late 1950s, "about half of the new houses in the United States were produced by only 6 percent of the builders." In 2000, U.S. construction corporations made a combined net income of $1.34 trillion. Only 5.5 percent of those firms earned two-thirds of that income.[11] And as for holding land for four or five years during the planning and approvals process, only 1 percent of construction corporations nationally have assets of $10 million or greater—some much greater.[12] (This figure is calculated without including any of the one-man contractors that the 1997 U.S. Economic Census estimates at about 1.6 million, sharing an average annual income of about $54,000. It considers only business entities with paid employees.)

As the price of land rises, along with other costs of doing business, the developer passes these costs on to the consumer as higher housing costs and higher commercial space costs. We are paying, along with the developer, for the price of scarcer land and for the protections we need to preserve or wisely use our remaining land, to prevent air and water quality from further deterioration or prevent land use decisions that impoverish the community in other ways. Just as home building has begun to edge out other economic activity in certain regions of the country, it has begun to edge out its own interests—partly through practices that waste land and partly simply because of population growth—a phenomenon that fishermen on both coasts, facing depleted stock, can understand. As we have with so many other resources, we are entering the latter days of the land development frontier.

Homebuilding, like other economic activities before it, has moved westward, first to California, and now showing up as rapid growth in the suburbs around Phoenix and Las Vegas in recent decades. And yet, in the west, other limits to development demand our attention in the headlines. In these parched lands, seven states vie for the life-giving waters of a single river, the mighty Colorado. And as more and more people move to the southwest, they draw new battle lines in the wars over water.

CHAPTER 14

WATERING THE WEST

"You ain't never been in California?"
"No, we ain't."
"Well, don't take my word. Go see for yourself."
"Yeah," Tom said, "but a fella kind a likes to know what he's gettin' into."
"Well, if you truly wanta know; I'm a fella that's asked questions an' give her some thought. She's a nice country. But she was stole a long time ago . . ."

—John Steinbeck, *The Grapes of Wrath*

To an easterner, the most spectacular scenes of the West—snow-capped mountains, clear lakes, steep bluffs over the dazzling Pacific—hold a vertiginous grandeur, like looking in the face of God. As many pioneers found, the cozy green pastoral quilt of the eastern landscape prepares one for neither the variety of western scenery nor the dryness of much of the West. The sagebrush hills, the deserts, the treeless rangelands of brown grasses suggest profoundly different conditions for life. The very air seems thirsty, lifting moisture from the skin. The Southwest, in particular, has a climate mostly inhospitable to eastern molds and mildews but friendly to fires.

And yet, California, home of Death Valley, the Mojave Desert, and the Colorado Desert, home to some of the driest land in the Union, grows a quarter of the nation's agricultural produce. This cornucopia grows on land that, under natural conditions, is so dry that some of it rarely gets more than three inches of rainfall a year—nowhere near the twenty-two inches needed to sustain crops. Yet, in 2003, California reaped more money in "cash receipts from farm marketings" than any other state—nearly thirty billion dollars—and almost twice as much as Texas, the next state after California in agricultural earnings.[1]

Between California's scenic coast and the majestic mountains of its eastern border lies the ancient floor of one or more primordial seas, where wondering explorers have even found ancient oyster beds.[2] This central floor is called a valley in California although it does not resemble a traditional valley. From the middle of its seemingly limitless flatness one cannot see the ridges on either side that were once shores of an ancient sea. One sees only desert. These vast dirt floors where most of California's crops are grown receive an average rainfall of between 2.75 (in the Imperial Valley) and 10 inches annually (in the northern San Joaquin Valley).

Despite an occasional drenching, such as in 2005, Californians cannot count on rainfall for their water. Instead, snowmelts from the distant hills and mountains, mainly in other states, run in rivers to the Pacific through the dusty ground. Early settlers took drinking water and then water for crops from these rivers, experimenting with irrigation. Over the decades, some of these rivers and lakes have been drained dry by irrigation canals for agriculture, killing off both the water system and the wildlife dependent on it.

California's quest for water built aqueducts, dams, wells, canals, and irrigation projects across the state to support burgeoning cities. California towns and developers sought water the way eastern and midwestern towns had sought the commercial lifelines of canals, roads, and railroads. Water fueled dreams of agricultural empires in the desert. Speculators sought to match up agricultural land or urban populations with distant water sources. The quest for water led speculators and governments to consider the question of water rights and to reevaluate property rights in a region where land without a water supply other than rain was worthless.

The current headlines focus on the questions of where water will come from, who will own it or transport it, what will happen to the places from which it is taken, and how much of the population's needs the water will meet and for how long. The headlines document tensions between farmers who need the water to raise crops and officials of growing cities who need to maintain a supply of drinking water and standards of public health, and between different regions and states fighting for the same water.

ORIGINS OF THE HEADLINES

In dry Southern California, both Los Angeles and San Diego are pressed for water and both are growing, along with many smaller cities and towns. In fact, California has been adding roughly a million people to its population every eighteen months. When the sun rises on California, it still shines on a promised land of mild climate.

Since their beginnings, both Los Angeles and San Diego have been pressed by their growing populations toward titanic water projects. But while Los Angeles stole the public attention in *Chinatown*, and with many subsequent headlines concerning its long search for water, San Diego gets many of today's headlines. San Diego's water war is with the farmers of Imperial Valley, once part of San Diego County.

From the beginning, San Diego was fair and dry. Her climate smiles alike on vacationers in fashionable resorts and the homeless who comfortably walk her streets. And her calm harbor has always captured the sailor's heart, as it did Richard Henry Dana's when he arrived on the brig *Pilgrim* from Boston.

Dana had abandoned his studies at Harvard in 1834 to rest his failing eyesight. He sought a complete change by shipping out to California. A sensitive observer, even without keen eyesight, Dana translated his experiences into a popular book, *Two Years Before the Mast.* In it, he described the restful qualities of San Diego's harbor and the primitive stage of San Diego's development in pre–gold rush California.

> A chain of high hills, beginning at the point . . . protected the harbor on the north and west, and ran off into the interior as far as the eye could reach. On the other sides, the land was low, and green, but without trees. . . . There was no town in sight, but on the smooth sand beach abreast, and within a cable's length of which three vessels lay moored, were four large houses—built of rough boards and looking like the great barns in which ice is stored on the borders of the large ponds near Boston—with piles

> of hides standing round them, and men in red shirts and large straw hats walking in and out of the doors. These were the hide houses.

Dana later explains in part where the hides originate. On the sailors' day of liberty, they ride about three miles inland to find the fort, or presidio, and the small village huddled at its base, consisting of about "forty dark-brown-looking huts, or houses," peopled partly by the soldiers and their Indian wives, "and two larger ones, plastered." The crew then ride another three miles to find the more "striking" edifices of the mission and its square, occupied by wealthy priests, who are served by Indians whom Dana calls virtually enslaved, living in "thirty small huts built of straw and of the branches of trees." These Indians maintain the vast cattle operation, or rancho, over which the mission presides and from which many of the hides come. Between 1823 and the Mexican-American War of 1846–48, Mexico granted land amounting to about 948 square miles in this part of California to thirty-three separate owners, averaging thousands of acres each.

The mission-rancho Dana describes is reminiscent of the plantations of the colonial South, in which all wealth lay with the landowners while labor huddled empty-handed, making local economies poor and slow to develop. A related syndrome characterizes centers of agribusiness throughout the West and Midwest, including North Dakota, Iowa, and many parts of the dry valleys of modern California. The ranching industry that Dana observed was to die out in the drought of 1862–65, but the joint rise of speculation and agribusiness in California gradually replicated the old elite pattern of landholding with a new industry.

By 1850, with the gold rush swelling San Francisco's population by thousands of newcomers annually and California attaining statehood, San Diego still held only 650 people. But after the establishment of stagecoach routes to California in the 1850s and then the completion of the transcontinental railroad in 1868–69, a small boom reached San Diego. With 2,000 resident in the growing city, the San Diego Water Company formed to expand the water supply. In a full page in the city's 1874 directory, the company solicitously describes the technology that allowed it to solve San Diego's water problems by drilling two artesian wells, three hundred feet deep, and building two deep reservoirs—a proud feat of nineteenth-century engineering.

It was two decades later, in the 1890s, that the mayor of Los Angeles first thought of fleecing his own citizens with the landgrab on which *Chinatown* was based. Mayor Fred Eaton proposed buying the neighboring Owens Valley in order to use its water for Los Angeles. When water officials did not immediately agree, he bought it himself. City officials eventually had to buy the land from Eaton, who decided to make a considerable profit on the land. Meanwhile, the deal left the farmers of the Owens Valley high and dry as their water was pumped away to slake the thirst of Los Angelenos. Further landgrabs occurred a decade later when several wealthy industrialists acted on a tip from a water board member to buy land in the San Fernando Valley that was to receive new irrigation water from a new federally funded aqueduct. California sunshine turned water to gold.

In San Diego, too, public and private enterprise continued to build reservoirs and aqueducts to bring water from farther and farther afield to support the city's expansion until, in 1901, the city's eighteen thousand or more residents voted to purchase and manage their own water supply. The city bought the old artesian wells and eventually the San Diego "flume" and related reservoirs.

Like Los Angeles to the north, San Diego continued to search out and engineer additional water supplies until eventually its insatiable water infrastructure crossed the entire state. Southern California's Metropolitan Water District required federal grants and loans to complete the mammoth infrastructure. In the fall of 1944, with construction on the infrastructure not quite complete, San Diego faced a water supply crisis. President Roosevelt appointed a "Committee on the San Diego Water Problems." Shortly thereafter, the president approved the report to the U.S. Senate and "advised that he had directed the immediate construction, by the Federal Government, of an aqueduct connecting the Colorado River aqueduct of the MWD with the water system of San Diego at its San Vicente reservoir."[3] Around the end of World War II, after years of construction on dams, reservoirs, aqueducts, and pipelines, San Diego, Los Angeles, and eventually 135 other cities and towns in Southern California began receiving Colorado River water.

A NEW BEHEMOTH: CALIFORNIA AGRICULTURE

The spontaneous growth of coastal cities and efforts to meet their growing water needs differed markedly from the development of other parts of California and its large, thirsty agricultural industry. Historically, schemes to profit from the sale of land or the watering of vast tracts of dry soil for farming abounded. And the business of real estate speculation and hunger for profit permeated the cities. In the words of pioneer and missionary James S. Brown, who arrived in San Francisco in 1853, "It seemed to me that everyone was seeking his own gain, regardless of his fellow-man."[4]

Thanks to extravagant federal loans and land grants, the railroads were the largest landowners in California, and they promoted their land heavily. The Central Pacific Railroad, part of the transcontinental route that brought the population boom to California, owned vast chunks of the state toward the end of the nineteenth century, such as the checkerboard of roughly fifteen hundred square miles—or roughly a million acres—around beautiful Lake Tahoe. The railroads reaped great profits by selling the land and also heaps of resentment from displaced settlers and other victims of their wealth and power. That resentment erupted at least once into gunfire—at Mussel Slough in King's County—between settlers facing eviction and representatives of the railroad to whom the railroad had coincidentally sold the land. Frank Norris made the railroad's oppressive greed the subject of his popular turn-of-the-century novel, *The Octopus*, taking its title from the local nickname for the railroad.

But the Central Pacific Railroad and its affiliates were not the only opportunists in California. The new commercial potential of coast-to-coast rail transport attracted opportunists by the hundreds. The new railroad connections proved a primary catalyst of California's economic development. Earlier development impetus had come from the fur and hide trade, from the gold rush of 1848–49, and from early experiments with irrigation that spread to California by the 1850s. But when railroads successfully carried refrigerated California produce to markets farther east, the profits and importance of California's agricultural sector soared.

Finally, even later, when very little agricultural land was left to take up across the nation, the technology needed for large-scale water management projects developed, along with the political will to implement them. All these factors combined to support the development of agriculture as a vital industry in California. And all these factors inspired speculative schemes.

As always, luck in speculation varied. Irrigation in California's Central Valley began shortly after the gold rush. And near the Arizona border, Thomas H. Blythe became the first farmer in California to use Colorado River water to irrigate his crops in the Palo Verde Valley in 1956. Blythe prospered, and by the 1870s he was selling Palo Verde land to San Franciscans. The town of Blythe in Palo Verde bears his name. He filed the first California claim on Colorado River water in 1877 to irrigate forty thousand acres of land.[5]

Meanwhile, in the Imperial Valley, some dreamers and schemers who envisioned the valley's lush future productivity, on the other hand, died or went bankrupt before their much more ambitious project could be realized. It took the combined financial power of the U.S. government and the mighty Southern Pacific Railroad (an affiliate of Central Pacific) to give that fantasy substance.

UNCLE SAM TO THE RESCUE

The natural marriage of land and water in the eastern United States, replicating that of Europe, left Congress unprepared for the large expanse of western lands that were useless without water.[6] But the shrinking supply of well-watered farmland in the East and Midwest, along with the growing U.S. population, made speculation in the development of arid and semiarid lands inevitable. As urban populations of the West grew, and as irrigation gained popularity, and competition for water increased, Congress was obliged to attend to the distribution of a limited water supply.

The question of water rights fell to the federal government as part of the formulation of property rights and division of resources in a period of increasing competition among individuals, localities, and states for western water. Water distribution also fell to the federal government because enormous sums of public money were needed to engineer and build the systems that would divert, hold, convey, and distribute water: enormous dams, aqueducts, and canals and vast irrigation systems.

In 1902, Congress passed the National Reclamation Act, providing rights to irrigation water to owners of parcels of up to160 acres of land. With this act, Congress aimed, as in most of its land use policies of that era, to promote settlement of the country and the welfare of what Paul Schuster Taylor called "home-makers," by which he meant people who would live on and work their own land. "If it were not for the national irrigation act," President Theodore Roosevelt told an audience in California in 1903, "we would be about past the time when Uncle Sam could give every man a farm."

The programs of Uncle Sam, however, stood in rather frail contrast to the forces of development already marshaled in California. Investors in the Imperial Valley, for example, managed to construct enough of their new infrastructure—simplified by the below-sea-level elevations of most of the valley—to start Colorado River water flowing to the valley by 1901, a year before the 160-acre landownership limits of the National Reclamation Act went into effect. Although flooding and financial problems and changes of ownership marred the early years, the experiment eventually succeeded. Swatches of tender rich green cropland began to carpet the desert floor.

By 1905, nine to twelve thousand people, the majority of them Mexicans working for absentee owners, were growing 150,000 acres of crops in the Imperial Valley.[7] That fall, men completed grading the main street of the future Imperial County capital, El Centro. But that same year, the Colorado River overran the initial irrigation infrastructure, flooded the valley, and continued to wreak havoc over the next two years. President Roosevelt asked the Southern Pacific Railway Company to intervene with a large loan for repairs and rebuilding. In 1911 "the Imperial Irrigation District was organized under California law."[8] By 1916, the Imperial Irrigation District had acquired all of the interests in the valley from the Southern Pacific. The valley continued to grow with its mix of federal help, private utilities and other private investment, and large farms whose owners lived off-site. These farms seemed immune to the farm size requirements of the National Reclamation Act.

Federal engineers, however, had begun investigating the problem of harnessing the Colorado almost as soon as the National Reclamation Act was signed, and it was the Imperial Valley that spurred these investigations. The definitive federal report, "Problems of the Imperial Valley," came out in 1922, the year that seven states negotiated the Colorado River Compact. The report recommended damming the Colorado River at or near Boulder Canyon. The government planned to pay for the project with the profits from land sales, power sales, and minimal charges to farmers for irrigation water.

The federal government, acting under the auspices of the Bureau of Reclamation, now began an era of heavy spending on the construction of dams, reservoirs, canals, and aqueducts. The Hoover Dam built at Boulder Canyon in the early 1930s, which watered Imperial Valley and much of Southern California. In 1935, the government undertook the Central Valley Project, damming the Sacramento River to provide irrigation water for the highly productive San Joaquin Valley and potable water for San Francisco. And in 1956, the U.S. government added the Glen Canyon Dam to the lower Colorado.

Meanwhile, the search for water shaped not only California's history but her government. In 1925, the California legislature authorized the establishment of metropolitan water districts throughout the state, whose authorities would procure and distribute water among the residents under their aegises. For growing metropolitan areas, this was an increasingly ambitious and expensive undertaking, involving increasingly sophisticated and costly infrastructure that crossed increasingly large swaths of land to bring water from ever more distant sources.

GROWTH OF THE IMPERIAL VALLEY

To an outsider, the Imperial Valley may look like the middle of nowhere, but to 142,000 people, it looks like home. It has the feel of Richard Henry Dana's description of Mexican San Diego, with economic and social growth choked off, at that time, by the tight grip of wealthy priests. Though it does boast one college, its streets are lined with uninteresting single-story strip development, with gaps here and there, like the smiles of the very poor. And after school its libraries fill with Hispanic children. Hispanics make up nearly three-quarters of the population, with many of the farm owners still living outside the valley, as in the beginning. The median family income is about fifteen thousand dollars less than the U.S. average, and the number of high school and college graduates is lower.

Paul Schuster Taylor describes the early Imperial Valley in a 1973 article:

> By the late twenties Imperial Valley was not a traditional homogenous community of farmers working their own land, but a polarized, divided society. Operation and ownership of land were concentrated in a few hands; the laboring landless were numerous.

Taylor also writes that "By 1969, the size of the average irrigated farm in Imperial County was 494 acres, more than three times the 142-acre average in the State of California." By this time, western farmers were irrigating roughly sixty-four million acres of land.[9]

Paul Schuster Taylor was a specialist in dry, dusty ground. He is probably best known for his work with photographer Dorothea Lange, his second wife and the great passion of his life other than his work. They met when a California state agency hired the two of them to document the living conditions of the displaced rural poor during the Depression. Dorothea took now-famous research photographs and Paul took notes. They traipsed around the Dust Bowl together and, after completing their report, published *An American Exodus: A Record of Human Erosion in the Thirties*, which became the classic work on rural life during the Depression.

Taylor, who had grown up on a farm, never stopped considering the plight of the farmworker or the small farm owner. He became a farm labor economist. In 1943, the office of the secretary of the interior hired him as a consultant. This began, for Taylor, the last stage of his career—about thirty years during which he championed the National Reclamation Act and attempted to prevent abuses to it. He served the government for nearly a decade, during which time he saw more and more abuses of the system and more and more efforts to overturn the provision of the law that limited federal irrigation subsidies to farms of 160 acres or less, known as the Excess Land Law. As larger plantations increasingly gained access to free or cheap federally supplied water, he left and tried to call attention to the abuses to the federal reclamation laws by publishing journal articles for the next twenty years.

Imperial Valley had always flouted the law. But pressure to actually overturn the law was mounting. The same year that Taylor began his consultancy, a California congressman proposed exempting his constituents from the law. In 1950, according to Taylor, Richard Nixon, a candidate for Congress from California, "quietly let it be known in the Imperial Valley, where landholdings were large, that he was against the excess land law." In the end, Taylor writes, the Korean War may have overshadowed this issue, but Nixon was elected.

California would take the largest share of the Colorado. And the Imperial Valley, with its early claim to the river, would take the lion's share of that. California's total annual allotment of the Colorado River is 4.4 million acre-feet. Of that, the Imperial Valley receives 3 million acre-feet annually, or about 978 billion gallons. Statewide, agriculture consumes about 80 percent of the water used in California.[10]

Today, according to the 2002 U.S. Census of Agriculture, the average California farm size is 346 acres, rather than the 142 acres mentioned by Taylor for 1969. But the real indication of the triumph of agribusiness in the state and in the modern U.S. economy lies in another figure. A tiny 13 percent of California's nearly eighty thousand farms take in 77 percent of farm income, that is, the market value of agricultural products sold plus government subsidy payments.

THE WATER WARS

And now we have set the stage for California's intensifying competition over water. Thanks to generous federal subsidies, farmers, mainly large-scale business owners, control a large share of water resources for California. They use 80 percent of California's water. But California's cities and towns are gaining population, and their thirst is growing. California, by the late decades of the twentieth century, was annually overdrawing its share of Colorado River water, using part of Nevada's and Arizona's allotments. But increasing development in Nevada and Arizona required those states to demand their full share.

When California failed to comply with federal requests to live within the limits of its agreement with other states, the government issued a deadline. When the deadline passed, in December 2002, and California continued to surpass its limits, the Interior Department announced its plan to cut California's water supply. San Diego's negotiations with the Imperial Valley farmers broke down more than once. But by October 2003, the farmers agreed to sell some of their water to San Diego. San Diego would initially pay the farmers $258 per acre-foot. (An acre-foot supplies about two households with water for about a year.) The farmers would continue to pay the government $15 to $20 per acre-foot.[11] They would let some of their fields lie fallow—and collect a tidy profit.

In 2005, farther north in California, the U.S. government quietly began signing contracts allowing Central Valley farmers—who have not nearly repaid the billion-plus dollar cost of the federal dams and irrigation works that supply them with cheap water—to sell water to Southern California. The farmers buy water at a rate of between $16 and $61 an acre-foot and will get to resell it for the average $500 per acre-foot that Southern California's urban water agencies now pay.[12] Since an acre-foot supplies roughly two households for a year, these agencies are paying approximately a thousand dollars a year for every four households in their jurisdiction.

In the meantime, over these last several years, numerous schemes to import water through arrangements with private companies and corporations have come and gone. The public is not eager to see its water supply in private hands, especially since the most ambitious of these arrangements, such as the failed plan of the agricultural giant Cadiz to store water in the Mojave Desert to ship to Los Angeles, would require federal assistance.

A flood of questions accompanies these new developments. Some experts argue that placing water in private hands and putting a higher price on it will stem waste and spur conservation. And yet, the question of owning water is so much more complicated than that one simple calculus.

We might ask what would happen if an unscrupulous company decided to price water out of sight. Anyone who finds that unlikely should be aware that Enron had entered the water business a few years before its demise. It was in 1998—not long before Enron made a shambles of California's energy supply and state budget in the summer of 2001 by manipulating its wholesale energy supply and thereby prices. Enron had created the Azurix Corporation to provide water and wastewater services to localities. It bought a public water utility in Delaware, successfully bid on water and wastewater "concessions" in South America, and brainstormed about acquiring water through schemes such as bidding to clean up Florida's Everglades in exchange for rights to water there. In 2000, Azurix launched its "water2water" Web site, created for trading water in much the same way that Enron shunted energy around the country. Azurix died with Enron, but other large private water schemes are thriving.

Other troubling questions arise concerning private ownership of water or the means to convey it. We need look no further than the other markets that determine the quality of our lives to examine one issue. In the case of all markets—for land, homes, higher education, medical care, commodities, and consumer goods of all kinds—the market makes, or rather the purveyors make, a particular good available to those who can pay for it. Those who can pay more generally get the best quality. Those who cannot pay get none. Should water be available only to the well off? How about good quality water—free of bacteria, lead, and other pollutants? Actually, this trend is already under way in America, with the better off having spring water delivered to their homes and the rest of the population taking their chances with tap water, the quality of which changes by locality.

Another question about private ownership of water concerns where it will come from. Will private owners decimate vast ecosystems by removing water from them to transfer it to a large, thirsty urban population? Actually, this has already been accomplished by public irrigation and water supply projects—all those dams, reservoirs, and aqueducts—but the public is at least able to monitor and plan such projects. What if they occurred outside the public spotlight, outside the purview of public monitors? This ecological damage will continue and grow as long as the U.S. economy and population grow, but it deserves careful observation and regulation.

And yet, in a way, these decisions are already made privately, by the developers who construct housing in arid regions, the businesses that locate there, and the government oversight—or lack thereof—that allows it. On January 9, 2005, the *New York Times Sunday Magazine* carried a five-page advertising section with the title "Luxury Homes and Estates: Desert Living." Although most of the landscapes shown in the photographs of the luxury desert housing development featured cacti—a welcome change from the long-promoted notion that arid climates will support green lawns—the ad also featured a luxuriant green golf course. And one assumes that the people who will live in these luxurious desert accommodations are as thirsty as anyone else. When existing supplies run low, as they already seem to have done, will the residents of these new developments demand, too, huge federal interventions to find water—a huge taxpayer subsidy for their drinking water? Or will they help to drive up the price of private water supplies as population continues to grow and water becomes scarcer? Or both?

The scarcity of western water brings up one more issue. With much of the populations of seven western states dependent on the Colorado River, the federal government must be hypervigilant concerning potential pollution of the river. The Environmental Protection Agency (EPA) was recently fighting a proposal by another federal agency—the Department of Energy—"to leave 12 million tons of radioactive waste next to the Colorado River near Moab, Utah."[13] If anyone wonders how scarce resources have to become to earn federal protection, the Colorado River offers an example.

And anyone who thinks that water from the eastern and midwestern United States will rescue the West should consider the recent headlines in those regions. Towns farther and farther from the Great Lakes seek access to Great Lakes water. This region already fears western requests for its water. Farther east, Maryland and Virginia fight over the use of Potomac River water. New York City owns over a hundred thousand acres around its upstate reservoirs and continues to buy more in order to forestall development that could pollute the city's drinking supply. The city recently sought agreements with communities around the reservoir to limit development—to prevent pollution of the reservoirs—in exchange for recreational use of the buffer zones around the reservoirs and for financial aid from New York City.[14] Populous southern Florida has asked northern Florida to send its water south—an idea northern Florida does not welcome. Florida has even considered expensive desalination plants to convert seawater to freshwater. And periodically, various localities around the east institute watering bans or other measures to limit water consumption during droughts.

California, with its sun, surf, movie studios, and Silicon Valley, may be the place that has come closest in American culture to the myth of the promised land. Its history as the birthplace of the Central Pacific Railroad, the first large corporation on a truly modern scale, and its thirty-four million people and their lifestyle make California a good place to face again the twin engines of economic and population growth, the demands they place on our resources, and the changes they bring to our democratic institutions.

Now that we are back in California, it is time once again to evoke the early observers of our democracy. Once again, we position them in this sunny promised land, with a population of 34 million, most of whom clamor for California's warm-weather, good-times lifestyle and a very small percentage of whom, subsidized by the government, use large areas of land and large allotments of water to produce food for the nation.

Rubbing shoulders with thirty-four million people, what would our committee—Malthus, Macaulay, Gallatin, and de Tocqueville—make of Eden and its serpents? To familiarize them with these issues, we would have to acquaint them with the past several years of headlines concerning California. To assess the effects of population growth, they would have to learn that a majority of our least affordable housing markets are in California, the nation's most populous state. They would learn that Californians compete for water, that the cost of drinking water is high and rising, and that the cost of water for agriculture is federally subsidized. They would learn that energy, too, is shipped in at high costs and that high demand has made California vulnerable to the price-fixing practices of corporations such as Enron. They would learn that wealthier commuters can pay to drive in the so-called Lexus lanes—with electronically collected tolls of about eleven dollars per round trip—to save some of the ninety-three hours that the average Angeleno spends stuck in traffic.[15] Even the cost of the tiny piece of real estate occupied by a car on the highway is rising steadily out of sight and sorting people by class.

We could count on Malthus to consider the condition of the people—a high average standard of living masking growing inequality. Macaulay—who said that America's institutions would not be tested until population grew and land supply diminished—would look at the condition of democracy itself as would de Tocqueville. What would they make of the economic power and enormous political clout of large corporations consuming vast resources? What would they make of government's response—or lack of response—to the problems created by three hundred million people performing the simple, daily acts of drinking water, washing, eating, or driving a car? And what would they make of the way corporate and industry influence has compromised those responses? It is our responses to these problems that will show how our democratic institutions are surviving this challenge.

Macaulay and de Tocqueville—and all the rest of us who would like to ask and answer this question of how our democratic institutions are surviving the stresses of population growth—must measure how we protect the resources our society needs and how we apportion them. How shall we govern the competition for land and water needed by both urban populations and farmers? How will individual rights fare against the ever more complex needs of our ever-growing community? To what extent will a person's rights to resources be based on his or her wealth? Does a promised land attract too many of the greedy and the lazy, and, if so, how do we govern these qualities in ourselves? We must begin to answer these questions. For as our population continues to grow, it will take more than scholarly observers to answer these questions. It will take the wisdom of Solomon and the patience of Job.

SECTION 5

Americans and Their Land

Men have forgotten how full clear and deep
The Yellowstone moved on the gravel and grass grew
When the land lay waiting for her westward people!
—Archibald MacLeish, "Empire Builders"

CHAPTER 15

AMERICANS AND THEIR LAND

Man is the intelligence of his soil.

—Wallace Stevens, "The Comedian as the Letter C"

What is the relationship between people and their land? Perhaps words are inadequate to describe the profound and evolutionary character of the bonds between people and their natural surroundings. Our relationship to our home environment—the need to navigate it, predict its cycles and laws, find its most promising sources of food and shelter—gave rise to the sciences, from astronomy and physics to biology and chemistry. The human relationship to nature also inspired early art forms, from sculpture to storytelling and myriad crafts as well as mystical marriages of arts and sciences that survive in enigmatic monuments such as Stonehenge and daily activities as humble and familiar as cooking. The whole human consciousness has been shaped by this relationship in ways both deeply pragmatic and intangible.

Although we have so far considered mainly the economic, political, and social benefits of land and other resources that Americans enjoy, there are other, less quantifiable aspects of our relationship to nature that influence our daily approach to our surroundings. These less tangible aspects of our relationship to the land deserve a moment of our consideration if we are to understand the deeply personal individual experience of the land that, in aggregate, makes up our collective relationship to that land.

Until recently, researchers had seldom attempted to document our noneconomic ties to the land. We know that "open space" raises the real estate values of surrounding properties. We know that, in the last election, motions to conserve land were placed on 161 different ballots across the nation and that "120—or 75 percent—were approved by voters" to the tune of nearly four billion dollars "dedicated to land conservation."[1] According to the Trust for Public Land, as quoted in a 2003 *New York Times* article, voters over the last five years "have approved more than $23 billion in public money across the country to preserve land from development."[2]

We know that people seek out natural settings for their vacations. In 1999, the national parks received 287 million visitors, approximately the U.S. population at the time, though obviously this includes some repeat visitors rather than everyone in the nation. And now, thanks to recent research, we know that vacationers in natural surroundings come back with enhanced concentration, as shown on proofreading tests—a benefit not enjoyed by vacationers in urban surroundings.[3] So, in short, we know that nature has a value to people beyond its economic value. Conversely, we see that this intangible value also gives land some of its economic value. And we can also see that the average suburban yard, while emblematic of that human craving for nature, does not entirely satisfy it.

Far from the popular headlines, some investigation into these intangible benefits of nature has begun. Nancy M. Wells, a researcher in environmental psychology at Cornell University, dug into these studies before designing her own. She found research that showed adults and children benefiting in a variety of ways from contact with nature. Assorted studies showed greater "psychological well-being, superior cognitive functioning, fewer physical ailments, and speedier recovery from illness" among adults.[4] The effects of nature on children, which Wells researches, are even greater. Children have shown greater capacity for attention when they live within view of nature or go to schools surrounded by greenery, and children in rural areas show lower levels of stress.[5] Adolescent psychologists have recently added access to open space or nature to the list of things such as family dinners and sports that keep teenagers on the straight and narrow path, away from drugs and other risky behaviors.

Anyone who observes young children can see that nature is the most engrossing playground and learning laboratory available to them. The game of crossing the stream on rocks gives way to throwing in sticks to watch them ride the current and to quieter discoveries such as bird nests or novelties of plant life, all of which lead to a sense of wonder. And anyone who has grown up close to nature, observing the habits and life cycles of different species and their relationship to the great cycles of the earth's rotation and orbit, knows there is a link between learning and exposure to the natural world that has been little explored or assessed.

The very need to navigate the space around us helped to wire our brains, as Howard Gardner points out in *Frames of Mind.* Peter Freyd, professor of mathematics at the University of Pennsylvania, has made the observation, based on anecdotal evidence, that almost all of the most distinguished geometers, perhaps the top fifty, grew up in or near rugged terrain or in families who skied regularly. Their early experiences negotiating three-dimensional landscapes, Professor Freyd postulates, may underlie their superior ability to conceptualize in three dimensions. No doubt, human biology includes many such unrecognized or unexplored adaptations.

Before research and theories of this kind, the history of American arts and literature testified to the impact of nature on the American psyche. The Hudson River School of landscape painting arrived to glorify nature at the very moment nineteenth-century industrialization threatened to sully it. After the Civil War, Winslow Homer modernized the American landscape painting in extremely popular works. From the birth of the Union, American writers such as Cooper, Hawthorne, Melville, Thoreau, and Whitman used the natural world to illuminate human nature. Emerson gave the nation his transcendentalist views of the spiritual and therapeutic aspects of the human relationship to nature. Certain modern writers, such as Ernest Hemingway, and modern poets, such as Robert Frost, continued this examination of the intimacy between man and nature.

In particular, the impulse to exchange the imperfect society of man for the healing society of nature animates certain American children's classics. *Little House on the Prairie*, for example, opens with the family's decision to leave the Big Woods of Wisconsin, where more people are arriving every day. "Wild animals would not stay in a country with so many people. Pa did not like to stay either. . . . He liked to see the little fawns and their mothers peeking out at him from the shadowy woods." In *The Yearling*, Marjorie Kinnan Rawlings describes with greater lyricism a father's attraction to the restorative power of nature, along with the healing reverence that nature sometimes calls forth from the human heart: "The peace of the vast aloof scrub had drawn him with the beneficence of its silence."[6]

The great westward sweep of Americans bent on capturing land, though we have considered it largely in economic terms, had a psychological component as well. Emerson called it the hope of a better society. Others have stressed the impulse to leave a flawed society or the hope of a lack of society.

Though we cannot quantify these things, we must acknowledge that some of these intangible benefits of nature underlay the homestead movement and now underlie the American Dream that shapes our landscape and economy. They also underlie the tug-of-war between property rights advocates and conservationists. The single-family homeowner hopes to procure a small piece of nature's beneficence *and* to attain financial security. The conservationist hopes to preserve enough of larger nature's beneficence for everyone to share. And, of course, the single-family homeowner and conservationist are often one and the same person.

And that brings us to the limitations of the individual's hallowed relationship with the land. First, the needs that motivate the individual's approach to land—both economic and psychological—do not encompass the larger picture. The individual often fails to see or consider either the resources remaining or the larger society that must share them. As poignant as we may find the individual hunger for both a connection to the land and the economic benefits that determine an individual's comfort, quality of life, and status in our society, we must admit this limitation. The individual rarely sees the whole picture and, even if he or she did, would sometimes act selfishly for his or her own benefit while harming his or her neighbors. To govern our use of land, we require a collective vision of the best uses for our resources and the best ways to share what remains to us.

Second, the individual's relationship to the land in America has been distorted by personal fantasy, sometimes shared as collective fantasies and often sold to us in a stream of advertising that shows happy, glowing people with luxurious, shiny products such as cars. Personal fantasy motivated the pioneer just as it motivates the buyer of a single-family home—a home designed as a small estate. Collective fantasies motivated the construction of the nineteenth-century City Beautiful movement and the national expositions that accompanied it, such as the 1893 Columbian Exposition in Chicago, which featured the "White City." When slums and the noxious effects of industry defeated the dream of beautiful, cultured cities, fantasy fueled the escape to the suburbs and the increase in speculation that captured the American imagination with its promise of wealth. Developers now capitalize on these fantasies in the layout, architecture, and names of the subdivisions and shopping centers they build—as do manufactures of consumer goods and the companies that advertise those goods. And we see now more than ever the individual retreat from citizenship to fantasy fiefdom as we take shelter from inconvenient realities in entertainment and material comforts.

Finally, we, as individuals, must consider our collective relationship to the land because it is collective systems—economic and political—that govern our individual access to land, resources, and all aspects of nature. And that access is diminishing. Americans see the landscape individually, but we design it together. Our invisible collaborations—in the ways we invest, in government policies, and in the daily choices of lifestyle that make up the American Dream—create the landscape we see and govern the share of land and resources that each of us may receive.

Ray and Harriet, a husband-wife team of dairy farmers, learned this lesson well during their decades of farming. Their experiences make a useful guide to the ways in which society—its economic and political structures—operates on our individual desires and economic productivity. Ray and Harriet now manage a town-owned educational farm in Lincoln, Massachusetts, but they began farming in Connecticut in the 1960s and later moved to upstate New York. They love farming, love the countryside, and love animals. But gradually, despite their best efforts, their dairy farm in upstate New York went the way of the Edsel, the LP, and the typewriter. In 2000, they sold their farm and moved to their present situation.

THE BALLAD OF RAY AND HARRIET

The story of Ray and Harriet begins with a prologue set in Vermont. In late June 2000—strawberry season in Vermont—my family and I were visiting the central part of that state. Vermont is dairy country, and the big headline in the *Burlington Free Press* on June 25 was "Vermont Farmers Fear Suiza Domination." According to the article, Suiza Foods had been buying up dairy processors around the country, particularly on the East Coast, where it had cornered, by that time, 50 to 70 percent of the New England milk market. It had bought twenty-five dairy processing companies in the previous seven years. It was now in a position to dictate terms to farmers.

Five years later, when I met Ray and Harriet, I heard about some of those terms. Ray and Harriet had farmed for decades, starting out in Connecticut and moving to upstate New York in 1976. They work hard. (A couple of our interviews and phone conversations occurred at times when I would rather have been asleep.) They had no interest in getting rich.

After moving to New York, Ray and Harriet watched the dairy business decline for twenty-four years before they got out in 2000. A United States Department of Agriculture (USDA) article that year described the changes in the dairy industry that drove farmers like Ray and Harriet out of business.

> Economies of scale . . . have led to fewer and larger dairy marketing firms.
>
> Traditional dairy companies that manufactured and sold a full line of dairy products (fluid milk, ice cream, cream, cheese, butter, and canned milk) have disappeared from the scene. In the 1960s and 1970s, institutional investors—pension funds, mutual funds, and the like—favored conglomerates and companies that diversified into a variety of product lines.

That last line means that the money that individuals—average Americans and wealthier Americans—invest in mutual funds, pension funds, and other investment venues goes to support the trend of corporate conglomeration. Those corporate conglomerates exercise an economic power that overwhelms not only large and small competitors but the democratic process. Although corporations have been outlawed since 1907 from contributing directly to a candidate, they can and do contribute millions to either political party. They can create and fund their own political action committees. And individuals in the corporation can donate the individual maximum to any candidate, as long as they do not appear to orchestrate their efforts within the corporation. Through these and other lobbying tools, not to mention the power of their pervasiveness in the American economy, these large corporations and well-organized industries shape our lives and landscape.

The USDA article goes on to say that "there were six large U.S. proprietary dairy companies in 1998—Dean, Suiza, Leprino, Schreiber, Southern Dairy Group, and Dreyer's/Edy's."[7] Since then, half the companies merged: Suiza acquired both Dean and the Southern Dairy Group. Suiza rechristened itself Dean Foods and has several subsidiaries that oversee its many brands. In 2002, Dean Foods sales neared nine billion dollars. If you have never heard of Dean or Suiza, it is because they maintain most of the brand names they acquire, such as Arnold breads or Edy's ice cream. Chances are, you have bought their products.

In the late 1990s, the U.S. Justice Department initiated an antitrust action against Suiza in Kentucky when Suiza tried to buy its main competitor for the school milk contract there. Further investigations of its activities in other states followed. Suiza settled certain charges by agreeing to close a limited number of its plants, though none of them were in the Northeast. And in the same period, Suiza paid seven million dollars to settle charges of tax conspiracy.

In daily life on their farm, Ray and Harriet experienced the changes brought by agglomerating food companies and their growing hold on government. Ray recalls the U.S. Supreme Court decision of 1976 that canceled some provisions of the 1974 Campaign Reform Act, such as limits on campaign spending, as a pivotal moment.

"As farmers," Ray recalled, "what we saw after that was a lot of unfair rulings coming out of the USDA." According to Ray, representatives of their milk cooperative went to meet with their congressman in Washington in 1997, the year that Suiza bought up Garelick Farms, the Northeast processor to whom they were selling milk. They were told that the congressman couldn't see them. They were offered instead an aide who had had an agriculture course in college. Then they saw their congressman leaving for lunch with a representative of the Kraft Corporation.

Ray gave examples of rulings from the USDA. "We got a letter from Suiza in 1999," recalls Ray. "It said, 'We have just received permission from the USDA to charge you $1.50 per hundredweight of milk to haul your milk from New York to Boston.' Then when gas went up, they charged more. When gas went down they didn't drop the price." During their years of farming, Ray says, milk prices paid to farmers ranged from about twelve dollars per hundredweight (one hundred pounds of milk) to about seventeen dollars. With a herd of about sixty cows, and with most of them giving nearly sixty pounds of milk daily, Ray and Harriet could ship about six thousand pounds of milk every other day—or sixty hundredweight—for a weekly income, after freight charges, of between twenty-two hundred and thirty-two hundred dollars to maintain the farm, the cows, and their family.

Ray and Harriet still remember in loving detail the care they gave their herd. "For a while we had the highest-producing herd in Colombia County, New York, and then in Chenango County. Our cows were long-lived, too. We fed them natural feeds. They went out to pasture from May to September. They had monthly visits from the vet, and they had no leg or hoof problems. We gave them a two-month rest from milking before we bred them. And the big corporate farms bought their cows from us. Now that all the small farms are going out of business, the large dairy farms have a hard time finding stock. They don't breed cows. They used to buy from smaller farms like ours that took better care of the cows and took time to breed them."

Ray explained that "Corporate cows get shots to increase their milk production every ten days. They add protein and sugar to their feed—molasses or chocolate syrup. Those corporate farms just use up the cow and in about three years, the cow dies." But the megacorporations like Suiza and Kraft that process milk rely on the corporate farm. Corporate farms can achieve the same economies of scale that megaprocessors can. They thereby achieve the lower prices that make their product attractive to megaprocessors like Suiza. And that's what the big food processors are looking for.

"The distributor, Suiza, constantly tried to get lower prices out of the farmers," Ray continued. "The dairy industry never had subsidies, but the big food corporations lobbied for subsidies to dairy farmers so that the processors like Suiza could pay less to farmers, and they got them. We saw it getting worse and worse, so we got out."

In 1997, with dairy processors swelling and dictating prices, the farmers of the northeastern dairy states fought for and won the northeast Dairy Compact. The compact required that dairy processors make payments to dairy farmers when milk prices dropped below a certain price. This seemed fair because processors, notoriously, do not pass the price savings on to consumers when milk prices are low and do not pass the profits on to dairy farmers when milk prices are high. Instead, savings and profits always go to the processor. But in 2001, the compact expired. It had two enemies: dairy farmers of the Midwest who wanted to sell in the Northeast and dairy processors. Suiza-Dean happened to be headquartered in Texas and happened to have donated generously to President Bush's election campaign. Although when the compact expired, representatives of the northeastern dairy states offered a national dairy compact that appealed to midwestern dairy farmers, the vote in Congress broke down along party lines, with Republicans supporting a federal subsidy, delivering big savings to Suiza-Dean and other dairy processors, and handing the bill to American taxpayers. This is how our democratic institutions fare as population grows and powerful corporations and industries gain influence over those institutions.

Ray was right. It has gotten worse. According to a 2001 *New York Times* article, "the dairy industry is moving westward, to states like Arizona, California, Idaho and New Mexico, where large-scale agriculture is being adopted more rapidly."[8] The migrating industry leaves behind not only the Northeast but the Midwest, including the cheese state of Wisconsin. The article quotes Todd J. Duvick, a food analyst at Bank of America. Bank of America is at the top of the bank food chain, in which large banks have gobbled up smaller banks almost as hungrily as Suiza has gobbled up dairy processors. Duvick's assessment reflects the outlook within large corporations like his: "It's just part of the problem of the industrialization of agriculture," he says. "States in the Midwest have laws against corporate agriculture. There's a mentality of preserving the family farm. But in the process they're losing an entire industry."

The midwestern governments trying to preserve the family farm call to mind the loss of other community-minded land policies in

American history, such as the Puritans' land-sharing, community-minded values that dissolved in the individualist economic milieu of the westward-spreading nation. In the simplest terms, mammoth corporations now acquire power over entire regions of land and entire subpopulations—such as the farmers or blue-collar workers of the region—without actually buying the land.

Suiza-Dean is tiny compared to the tobacco and cheese giant Phillip Morris/Kraft, with revenues of more than fifty-seven billion dollars in 1999, or Nestle, now making ice cream in South America and with total annual sales of nearly fifty-two billion dollars in the same year. A number of economic factors swell the giant conglomerates, including globalization and economies of scale. Consolidation in one industry encourages consolidation in others. Large restaurant and supermarket chains, the clients for the vast food conglomerates, now make "centralized purchasing decisions." They place vast orders and prefer "standardized products."[9] The banking industry has also consolidated, creating larger and larger investors that seek secure investments in the form of large, diversified companies.

The shareholder's lack of accountability for his or her investments is the final blow. Shareholding is the ultimate popularization of speculation. Although it benefits mainly the wealthy, shareholding gives the hope of capital gains—having one's money make money—to more Americans. And Americans do not call on the shareholder to invest his or her money in socially or environmentally responsible ways—in corporations that practice responsible conduct toward society in their employment and environmental practices. We ask the shareholder only to consider the possibility of his or her own profits in choosing investments. And we ask the corporation to consider, as its main legal obligation, its duty to safeguard shareholder profits, not the well-being of society. This is a boon to large conglomerates.

According to one article, Suiza was handling "70 to 80 percent of the fluid milk in the Northeast" by 2002, after its merger with Dean. Between October 2001, about the time Suiza acquired Dean, and October 2002, the wholesale price of milk paid to farmers dropped from seventeen dollars per hundredweight of milk (minus freight charges) to eleven dollars per hundredweight (minus freight charges). The price of milk in grocery stores did not change during that time, and Dean Foods (the former Suiza) appeared to be profiting from its vicelike grip on the dairy industry, inspiring one farmer to ship manure to the processor instead of milk.[10]

Farmers are not the only people at the mercy of these mammoth enterprises. For example, in 2002, Dean employed about twenty-seven thousand minimum-wage workers.[11] The food industry in general uses vast armies of minimum-wage workers. According to Eric Schlosser, "The restaurant industry is now America's largest private employee and it pays some of the lowest wages."[12] Barbara Ehrenreich documented the lifestyle of a low-paid waitress in *Nickel and Dimed*, and it is not the way most of us would like to live.

Through monopolies on jobs, large conglomerates further contribute to the rising inequality in this country. And American corporations have raised the boss-worker gap in compensation to obscene levels. While in Japan executives typically earn 10 times more than the employee average, German executives 11, French executives 16, and Canadian executives 21, American executives typically earn 531 times more than the employee average in their company.[13] This emphasis on the role of the CEO in charting a company's success seems to have contributed to the rash of corporate frauds in the last decade.[14]

This widening gulf in employment prospects, exacerbated by our increasing technological sophistication, makes education and technical training more important than ever, just when the price of a college education is pulling further and further out of reach of the middle and lower classes.[15] Education has been an entrée to the middle class since colleges of law and medicine spring up in England in the sixteenth century. But at the moment, competition for schools is not only financial; it has enslaved children from their preschool years on as they compete, under the tutelage of their parents, for those spaces at the top, carrying workloads that child labor laws once banned.

The one thing Americans at the top and bottom seem to have in common is that we are all working longer hours. As my foreign-born mother-in-law observes, "Everyone thinks Americans are so rich, but nobody works harder for their money." In 2001, while other industrialized nations cut back slightly on their hours, Americans worked more than ever and more than workers in Japan, Canada, Britain, or Germany.[16]

These changes affect all avenues of our society. The continued demise of family farms such as Ray and Harriet's is only one example.

Remember that 550 acres was a large farm in Jonathan Hale's day and that his proximity to growing markets was a key to his prosperity. American population is still growing, causing milk and other markets to grow, but those markets are now controlled by mega-distributor-processors that make Ray and Harriet's sixty cows on similar acreage insufficient. We have all colluded in robbing families such as Ray and Harriet's of their independence, their chance at economic self-sufficiency and our own.

Almost every American industry has seen the trend toward consolidation and megacorporations, with increasingly powerful corporate or industry lobbies influencing the quality of our lives by influencing government policy. Corporations or industries direct public policy to serve their own needs, without much awareness on the part of the public. This means not only that the government rewards those who can afford to influence officials and public attitudes but also that we, the public, are paying industries and corporations, through the awarding of various advantages, to influence our government (and sometimes ourselves) against our own interests. In other words, we, the public, pay these powerful commercial interests to influence how we live, how we spend our money, how we use and apportion resources—and to do so in ways that profit those interests.

Since I began researching this book, corporate or industry lobbies have squelched health-care reform; squashed even cost-of-living increases in the minimum wage since 1997; influenced Food and Drug Administration (FDA) recommendations concerning food consumption and public health; passed the cost of recycling bottles on to the state and stores; swung local elections or influenced legislation that concerned development issues; convinced the Federal Communications Commission to "relax the nation's media ownership rules" to allow further consolidation of media control by a limited number of large corporations;[17] curtailed Americans' recourse to bankruptcy; achieved greater access to public lands for timbering and mining interests and greater access to national parks for snowmobiles; and removed regulatory obstacles to strip mining, to name just a few examples. The pharmaceutical industry recently meddled in the legislative process to limit the power of government and citizen to obtain reasonable prices on drugs, and some drug companies have also flouted government regulations that require them to publicize relevant details of studies revealing their drugs' effectiveness and side effects. And, of course, industries, notably the power industry, have continued to rob us of our resources through environmental contamination, supporting our illusion of luxury and convenience while poisoning the air we breathe.

This particular example has attracted more attention as discoveries of higher quantities of mercury in our food supply have raised alarms. Though mercury levels in fish have recently caused the FDA to rewrite some of its guidelines for human consumption of fish, the power industry, the chief source of mercury in our environment, has so far been able to avoid constraints on its mercury emissions commensurate with the public health crisis created by mercury, a carcinogen that causes birth defects. (Some pediatricians are aware that the tuna industry exerted pressure on the FDA not to ban tuna, which shows high levels of mercury. Although we could have protected both the tuna industry and the consumer by passing stronger antimercury legislation, instead the risk of eating lower-quality food was passed on to the consumer.) Also consider the lawsuits pressed by various groups of eastern state governments against the federal government for failing to protect the quality of their air from the power industry and other sources, in one case by failing to follow its (the federal government's) own oversight procedures.[18]

Think of Enron, two of whose officers pled guilty to manipulating the California energy market in the summer of 2001, when Enron created an artificial shortage that allowed it to bilk the California public—through prices charged to their utility companies—of astronomical sums for electric power. In 2004, still another executive, an officer of Houston-based Reliant Energy Services (a subsidiary of Reliant Resources), pled guilty to the same charges relating back to the California energy crisis of 2001.

And again in auto-swamped California, think of the Big Three automakers suing the state government, as they are currently doing, for passing auto emissions standards that would force carmakers to offer hybrid vehicles. The Big Three find this unfair even though Congress began subsidizing the automakers' research into hybrid and electric propulsion systems and vehicles in 1974 and, by the 1990s, was paying tens of millions of dollars annually to fund that research.[19] Meanwhile, Honda and Toyota have both come up with a marketable hybrid vehicle during that time, while American automakers have brought us the gas-guzzling SUV and the minivan and have promoted the pickup truck as a macho toy.

The subject of the car brings us to another aspect of industry political clout. Earlier chapters of this book dealt with the influence of the real estate and construction industries on our patterns of settlement and homeownership. The car, as others have pointed out, also helped to design our present landscape. Not only does the transportation sector consume the majority of petroleum used in the United States—half again as much we produce[20]—but the automobile also uses more land than we can comfortably afford, for roads, parking, and wasteful, low-density layouts. The car underwrites our sprawling pattern of settlement, a pattern that uses vast amounts of land, makes cars ever more indispensable, and demands the use of more and more fuel every year by people trying to navigate its growing morass.

And yet, like the subdivision, the car represents a kind of collusion between the public and industry. The consumer longs for luxury and convenience, and the home builder, the car manufacturer, and other manufacturers package those qualities in a salable form. No agency oversees the resources that will be consumed by the manufacture of homes, automobiles, fuels, or any consumer goods or the environmental impact they will have when they become part of the lifestyle of three hundred million U.S. residents and an earthmoving force in shaping the American landscape.

Since air pollution was found to threaten public health, auto emissions have come under greater scrutiny for a few decades,[21] but the insidious forces—including our own proclivities—that have gradually increased our dependence on the automobile and thereby fossil fuels and particularly foreign fuels go unexamined. That lack of examination is a matter of public awareness and public will. The point at which public scrutiny falls short is the point at which we should start blaming ourselves instead of corporations. And since we each, as individuals, collaborate in sustaining these systems that use, abuse, and apportion our natural wealth, it is time to talk about how we can do a better job.

THE GREAT TASKS

Let's look once again at the effects of greater numbers of people in America sharing our resources. Land and housing prices rise. Our access to natural resources is more limited. Common resources such as air and water become degraded. Despite multiplying convenience and luxuries, the quality of our daily lives declines in certain ways: we work longer hours, for example, and spend more time in our cars commuting. Conflicts over the use of land increase, and increase in bitterness. Our economy grows larger, more complex, and apparently more productive, and its industries and corporations grow too, gaining sway over our institutions and agencies of government. Inequality rises. Technology expands for the increasingly difficult and complex harvest of resources, while its deleterious effects on common resources go largely unmonitored, and a limited number of people profit from the use of these technologies.

America's task in the twenty-first century is to address and redress these trends. Our collective mission is to steward our resources, for our own well-being and that of our children and their children, and to apportion remaining resources equitably among the large and growing population of a just society. The goal of equality is to allow as many Americans as possible to achieve their potential in order to make meaningful contributions to our society. It is a goal that benefits all of us—a breathtakingly desirable goal, and one against which economic forces militate. Our first step, therefore, as a few lawmakers have realized, is to reclaim our government from the undue influence of those economic forces—the lobbying of industries and corporations that shape the spending of government and every aspect of American's daily lives, including wages, housing, health care, job safety, transportation, fuel and energy supply costs, investment, savings, and credit rules. The government, in turn, must shift its support for commercial enterprise and the innovation it brings toward the goals of resource protection and social justice.

Our over-reliance on market forces to distribute wealth has made the New World more like the older societies from which our ancestors came, in which wealth was political power and vice versa. (Meanwhile, western European societies have evolved, for the most part, more equitably than ours, partly because they have fewer inhibitions about taming market forces and funding social welfare.) Economic power must be guided by a collective vision of the community's good. We now know that promoting all forms of commerce to achieve national prosperity is too simple a recipe and that some kinds of economic activity have hidden social, political, and environmental costs, all of which are intertwined. Those who think that these goals require an unbearable increase in government regulation must consider that regulation and government agencies multiply anyway as our economic systems become more and more complex, and that it is the quality rather than the quantity of regulation that must change. Consider the *Economist*'s pronouncements on "America's Regulatory Mess" in its July 26 issue, 2003. The current administration, it pointed out, Republican, probusiness, and "deregulation-minded" had already "published more pages of rules and regulations last year than any previous administration in any previous year."

American government has always concerned itself with Americans' access to resources, the source of national productivity. This is, in many ways, a familiar role for government and a comfortable public trust.

Historically, the government intervened in the national economy when it distributed land, funded transportation projects, and diverted water on a large scale to permit farming. The passage of the 1880 Sherman Antitrust Act, the creation of the Securities Exchange Commission, the establishment of the FCC, the setting of interest rates by the Federal Reserve Board, and the historic regulation of the power industry are all examples of government monitoring of commerce in ways that promoted the success of the many and ostensibly limited monopoly, though those powers have recently weakened, apparently in the face of corporate or industrial lobbies.

In other ways, this province of public concern and government—that of stewarding resources and guarding equality—is different. We find ourselves in the role of sharing dwindling, rather than abundant, resources. And sharing comes hard to human nature.

Changes have already begun. We have regulated air and water quality for a few decades and to some small extent pesticide and toxin uses, as research has followed these pesticides through the food chain to the American dinner table. The EPA Superfund cleans up some toxic spills, although it has recently been hobbled by funding cuts. All of these regulations bear directly on public health, just as much as those of the FDA. Still, our efforts to keep poisons out of our lives frequently run aground on the demands of industry and the average American lifestyle.

There is no shortage of ideas for environmentally friendly economic alternatives and ideas for government programs. There is no shortage of efforts to mitigate inequality and of initiatives to better design communities or redistribute population by both the private and nonprofit sectors. These initiatives to help us adapt to our changing conditions spring up often but are equally often stymied or ignored. They are like weeds sprouting through cracks in a thick pavement. That pavement is the cemented economic power of large industries and corporate conglomerates over government, which we support through both political apathy and careless habits of financial investment. Only a lack of public support and investment prevents innovation from flourishing.

Ironically, industry's hold over Congress is so debilitating that, while senators and congressmen have proposed legislation to address relevant issues such as campaign finance, supports for community planning, incentives to redistribute population to less crowded, economically starved areas such as the plains, and other reforms, these efforts (such as the McCain-Feingold Act) have either been stymied or diluted.

Small grassroots efforts prove more inspiring. For example, when I had the good luck, a few years ago, to arrive with my family in Vermont at strawberry season and to read about Suiza Foods, we stayed at an inn owned by a charismatic restaurateur named Douglas Mack. Doug, a chef, had founded the Vermont Fresh Network. His network connected local farmers with local chefs who were interested in getting high-quality fresh produce. Other similar efforts are now flourishing, such as the much larger EcoTrust on the West Coast, and are recreating the local economies that have been declining and disappearing for over a century. They are behind the slight resurgence of small truck—or vegetable—farms. In certain locales, such farmers are able to reach the restaurant markets of nearby cities or their affluent residents. Through farmers' markets, in which farmers sell from the backs of their trucks, and through subscription memberships, farmers are able to reach these high-end markets. Of course, this is another example of the higher-quality life and resources that the affluent increasingly monopolize. But it helps the small farmer who is also a consumer and it supports in most cases a benign use of resources.

Similarly, a host of initiatives propose to reduce America's car dependence, and they come from many sources. New Urbanism and related planning reforms are private-sector attempts to reform sprawl and automobile use. Their compact walkable communities appeal to only about 10 percent of new home buyers, but that is not a bad start. These initiatives, inspiring as they are, are also small and inadequate to the task that a growing population forces on us with greater urgency. Denser housing is gradually increasing in response to population growth, but not fast enough. Further reforms must have government support. With enough investment, shared public spaces can be as beautiful as those in Paris, and a subway system can provide an experience comparable to a Disneyworld ride, like the Metrorail in Washington, D.C., when it was new.

Another private-sector inspiration is the Zipcar, the brainchild of Robin Chase, who conceived it as an urban dweller's alternative to car ownership: car sharing by the hour. Born in Boston and now catching on in New York and Washington, as well as on college campuses, the Zipcar encourages people to forego car ownership, relying instead on the Zipcar when needed but otherwise using mass transit and other transportation alternatives.

Saving the environment can be a source of economic growth: a whole slew of businesses now offer remediation systems for toxic spills and contamination of various kinds. The discovery that certain plants absorb and retain soil contaminants such as lead has created new business opportunities as well.

There is no end to the ways that conscientious government could shepherd our transition to an economy more beneficial to more Americans. It can aid in this conversion by redirecting subsidies and investments. It can assist the gradual migration of population from environmentally and socially ruinous enterprises to environmentally and socially friendly ones. It could provide deterrents and oversight through an agency like the FDA that would review proposed manufacturing processes for environmental impact—instead of leaving the discovery and clean-up of deadly toxins until too late—when both the monetary expense and the cost to public health have become very high. It can promote conservation through regulation and through the more elegant means of tax incentives. For example, Colorado and Arizona now assign tax breaks to purchasers of hybrid vehicles. Government can address resource conservation and rising inequality simultaneously by efforts such as increased public investment in mass transit. It could establish an affordable housing trust fund; like the one Congress is currently consdering, from a tiny percentage of the Freddie Mac and Fannie Mae profits. That is, government could once again nurture the interests of all Americans. But the success of all these efforts will depend on a long-term campaign of education and awareness.

THE END

Tocqueville was right: "Materialism is a grave danger in the population of a democracy." It has opened us to the seductions of wealth and those who cater to our fantasies of wealth for their own profit. It has engendered an economy that has greater power than the democratic mechanisms of government to sort resources and human beings to the fates that profit existing economic powers. And Lord Macaulay, though we hope mistaken in his distaste for democracy, rightly foretold the test of America's democracy that would come as people multiply and access to resources dwindles. The test is whether we can act collectively, whether we can understand together the invisible mechanisms that shape our lives, whether we can act with vision to support innovation and change, whether we can share resources equitably among ourselves.

Even Malthus should not be dismissed, as he sometimes is. Malthus is out of fashion because the starvation, war, and pestilence that he predicted would arrive with overpopulation have not arrived in the United States in the way that they plague more truly overpopulated nations. With our sophisticated technologies to harvest resources and to maintain our standard of living, and with our steadily increasing consumption of the rest of the world's resources, we still appear to be winning the race against scarcity. And the effects of degraded resources on public health have just begun to be understood. But if we look for changes that are not catastrophic but gradual—the gradual deterioration of our air, water, and land supply; the gradual increase in conflicts over those resources; the gradual co-opting of government by economic forces; the gradual rise in inequality; the gradual increase in wars and international tensions relating to oil and energy—we can see in the distance the shape of our evolving future. In our land of plenty, it is now the bottom half who find themselves priced out of certain resources, such as land and clean air, and dying for oil in the Middle East. They are, as usual, the canaries in the coal mine.

If we do nothing to save resources and counter the trend toward inequality, the market will continue to make changes for us. Industries, and the corporations within those industries, will continue to enlarge their economic and political power. They will continue to influence legislation to increase their advantages and leave the consumer and the worker—more and more Americans—less and less protected.

As metropolitan populations grow without significant changes in our transportation system, sprawl and oil consumption will continue to grow. Housing markets in the large, economically productive metropolitan areas where most Americans live will continue to intensify, and housing costs will gradually consume a greater and greater share of the American income.

The exurbs will continue to sprawl outward and to consume farmland, timberland, rangeland, and mineral lands. The government will continue to allow megacorporations to dictate to megafarms, compensating for the diminishing land supply with technological advances such as the cow-killing milk-production booster.

The economic patterns that have produced a minimum wage no one can live on will continue to advance as costs of living, on the other hand, continue to rise. Gradually, more and more Americans will find their lives, though full of luxuries compared to a century ago, threatened by stresses of longer work and commuting hours; lower-quality air, water, and land; and less access to nature. And because we consume the rest of the world's resources, because we colonize countries economically through corporations that ignore the welfare of the people and their resources in those countries, and because we offer so many poor-paying jobs in this country, we will continue to attract large numbers of immigrants as a lifeboat attracts the drowning.

The market can make all these decisions for us, or we can try to design the future for ourselves, to reclaim the blessings of democracy and the blessings of our common inheritance, as Jefferson called nature. We can try to design an economy that is greener and less dependent on population growth so that we can consider slowing our rate of population growth. We can legislate to combat inequality. We can change our settlement patterns and transportation habits.

All of the needed reforms—to direct our habits in a more resource-friendly way, helping Americans understand how their consumer choices, for example, limit the quality and quantity of the resources to which they have access; to release the corporate hold on government; to promote greater government investment in and oversight of the public good—begin with the task of awareness and education, to which this book is dedicated.

Somehow we must emerge from our affluence-based fantasies and look carefully together at the larger picture around us—a picture of changing conditions in both nature and society. And we must substitute the dream of engagement—confronting our problems to find solutions—for fantasies of escape. Collectively, we must heal our individual relationships with our land, in the ways that research is now beginning to suggest, as well as with our democratic institutions.

Part of this effort—the part that will change behavior and raise political awareness to demand reforms—will be an educational campaign to restore our understanding of our part in the natural world and the natural systems that sustain us as well as the man-made systems that apportion it. The study of those systems, the consumption of the resources that are part of those systems, and our own role in that consumption complements the current passion to reinvigorate science teaching in American schools and suggests that increased contact with nature should be part of that effort. If the government cannot begin this effort, nonprofit organizations and others will have to do so. Another part of our efforts must improve education regarding citizenship and our responsibility to work together to equitably distribute resources.

A profound understanding of the world from which we draw life must once again become part of our daily thinking and our sensibilities. We must try to reconstruct the unfathomable intimacy with our land that has been key to humankind's survival for millions of years—an understanding of what the land can offer, what it can bear, and how we can and cannot use it. And with that understanding, we want to remember the sense of community that forced earlier peoples to cooperate for survival, and in that spirit of cooperation and enlightened self-interest, we must further examine the share of resources that each of us receives and how that share is decided and delivered. We must strengthen our democratic institutions again against the powerful economic tides that threaten them. And we must live on our land as if we wanted to stay forever.

Appendix

Symptoms of Stress in America's Most Crowded Housing Markets (cities and towns), 2000

Place	People per square mile	Percent of homeowners	Cost of median unit	Percent living more than 1 per room	Percent of homeowners paying over 35% of income	Percent of homeowners paying over 29% of income	Percent of homeowners paying over 24% of income	Percent of renters paying over 34% of income	Percent of renters paying over 29% of income	Percent of renters paying over 24% of income
Jersey City, NJ	13,043.6	28	$125,000	13	27	35	46	30	37	47
NYC	8,158.7	30	211,900	15	27	33	42	34	41	50
Nassau Co., NY	4,655.0	80	242,000	4	29	32	43	32	39	50
Passaic, NJ	3,724.6	38	153,000	24	33	41	51	34	41	53
Orange Co., CA	3,605.6	61	270,000	16	23	32	44	33	42	54
Los Angeles	2,344.2	39	221,600	26	29	37	47	37	45	56
Bridgeport, CT	1,754.9	43	117,500	8	25	33	44	36	43	54
San Francisco	1,704.7	35	396,000	12	23	30	39	28	36	47
Boston	1,685.1	32	190,600	7	20	26	35	32	40	52
Stamford	1,682.7	57	362,300	7	30	36	46	31	39	51
U.S. average	79.6	68	119,600	6	16	22	35	30	37	48
5 most densely populated places	6,637.3	47	200,380	14	27	34	45	33	40	51
10 most densely populated places	4,235.8	42	246,015	13	28	34	42	33	40	52

Note: The U.S. Census groups cities differently in different tables, obtaining different densities for differently configured metropolitan areas. The information in this table was gathered from tables at the U.S. Census Web site, www.census.gov, beginning with "Geographic Comparison Table," GCT-PH1(one)-R. Population, Housing Units, Area, and Density (geographies ranked by total population): 2000, from Data Set: Census 2000 Summary File 1(one) (SF 1(one)) 100-Percent Data, as well as individual tables of population and housing characteristics for the cities listed.

ACKNOWLEDGMENTS

Spouses usually come last in acknowledgments, but I must first thank my husband, Alex Krieger, for his part in this book. For the last twenty years, we have shared a conversation about American culture and settlements, and over the past few years, Alex gallantly suffered through many early chapter drafts, offering invaluable criticism.

Other than my husband, my good fortune in the writing of this book could not have been greater than in having Sam Bass Warner Jr. for a mentor and friend in the development and early writing of this book. As a historian and human being of the first rank and an early believer in the book's potential, Sam had a tremendous influence on the outcome of the manuscript.

Special thanks are also due Richard P. McDonough, lion-hearted literary agent, who years ago gave critical early advice and eventually found enough substance in the first few chapters to donate his services to find the book a home. At that home, the University of Michigan Press, my editor, Raphael Allen, demonstrated excellent surgical instincts regarding the manuscript and provided helpful doses of support and encouragement—another stroke of good fortune for me. His assistant, Christy Byks, also offered helpful insights. Again at the University of Michigan, Dean Doug Kelbaugh of the Taubman Center gave a close early reading and sage advice that improved the manuscript.

Jerold Kayden gave generously of his planning law knowledge. Many others helpfully fielded questions and supplied information in that field and others. I thank Matt Keiffer, Patrick Taves, Nicholas

Retsinas, Roger Booth, and Sarah Klain. In Loudoun County, my thanks go to Bob Lazaro, Peggy Maio, John Nicholas, and Julie Pastor; in Mashpee, I thank Tom Fadula.

Harold Pollman and Ray and Harriet trusted me with their stories, which brought much-needed life to the book. I cannot thank them enough.

Scholars who generously took the time to answer queries from an unknown writer include Martin Shanahan of the University of South Australia, Peter Freyd of the University of Pennsylvania, and Nancy M. Wells of Cornell University.

Countless research librarians facilitated the investigations behind this book. I am grateful to the Digital Reference Team of the Library of Congress; a host of reference librarians at Widener Library, Harvard University; David Cobb, director of the Harvard College Library Map Collection; Brookline Public Library; Ann Whitlow and others at the Mashpee Historical Society; many at the Cleveland Public Library, its Map Collection, Special Collections, and Government Documents Department; Ann Sindelar and others at the Western Reserve Historical Society; Susie Hanson and Mary Burns at the Special Collections of the Kelvin Smith Library at Case Western Reserve University; Steve Axtman at the Special Collections of the Chester Fritz Library at the University of North Dakota; and those at the Special Collections of the San Diego Public Library, El Centro Public Library, and the archives of the Pioneers Museum in Imperial Valley, California.

In addition to the U.S. Census Bureau Web site, statistics came, over the years of research for this book, from the Boston branch of the Census Bureau, the Weldon Cooper Center of the University of Virginia, and, most frequently of all, from the university's Web site of historical statistics at the Geospatial and Statistical Data Center.

And since the book's thesis was shaped by current events, I would like to acknowledge the work of journalists whose articles in the *New York Times*, *Washington Post*, *Boston Globe*, *Los Angeles Times*, *Seattle Times*, *Burlington Free Press*, *Cleveland Plain Dealer*, and other newspapers informed the book.

For some bits of artistry, thanks to Radhika Bagai for the maps—and to Genevieve De Manio for the photo.

In spite of all this information and assistance, any errors in the book are, as usual, my own.

Finally the formative contributions of family: My children, Isara and Isaiah, provided real assistance and patience in various forms, all deeply appreciated. My mother, Barbara Stone Mackin, with her trick of finding wildflowers and berries, gave me a close relationship to nature, my first books of poetry, and the latest clippings on Loudoun County. And years ago, my father, the late Welden Fuller Mackin, moved his young family to a series of planned communities. There he raised us on theories of social criticism. In this way and through genetic contributions, he disposed me to expect order and reason in my surroundings. The lifelong puzzlement to which that expectation led eventually became this book.

NOTES

CHAPTER 1

1. Samuel Eliot Morison estimates that about one in forty New England families sent a student to Harvard in 1640. "This was a much higher proportion of educated men to the population than could be found in any part of England at the time. In the history of modern colonization, it is unprecedented." *Builders of the Bay Colony* (Boston: Houghton Mifflin, 1930), 184.

2. Scrip for free land was used to pay America's soldier from the Revolution through the Civil War. Even after World War II, veterans were rewarded with loans for the purchase of homes on easy terms.

3. David Hamer, *New Towns in the New World* (New York: Columbia University Press, 1990), 70–73. The term *growth* throughout this book refers to both economic and population growth unless otherwise specified.

4. Thomas Robert Malthus, *Population: The First Essay* (Ann Arbor: University of Michigan Press, 1959), 119.

5. This interpretation of Balboa's experience comes from Keats's poem "On First Looking into Chapman's Homer," where, scholars agree, Keats misnamed Balboa "Cortez."

6. This information is based on U.S. Census statistics.

7. Land use figures have been categorized differently over the years. These figures were arrived at by taking land use statistics for 1930 for the contiguous forty-eight states from the United States Department of Commerce's *Historical Statistics of the United States: Colonial Times to 1970*, Part I (Washington, DC: Government Printing Office, 1975) and comparing them to 1930 population figures for the contiguous forty-eight states, which revealed a ratio of .43 acres of land to each person. Figures from 1992 for land categorized as urban or special use (such as transportation) were taken from the 1992 USDA publication *Major Uses of Land in the United States* (Washington, DC: Government Printing Office, 1992) and compared against resident population figures for 1992 to reveal a ratio of 1.1 acres of urban land per person. Urban land figures do not include farm buildings, and it is not clear whether they include development in land still characterized as rural.

8. William Fulton, Rolf Pendall, Mai Nguyen, and Alicia Harrison, "Who Sprawls the Most?" *New England Planning*, July–August 2001, 1, published by the Massachusetts and Rhode Island chapters of the American Planning Association.

9. A 2001 study by the Brookings Institute found that the area of developed land around the nation's metropolitan areas increased by about twenty-five million acres from 1982 to 1997, an area (if assembled together) the size of Indiana. The authors summarized their findings as "Sprawl Study Shows Massive Land Appetite," *Growth/No Growth*, August 2001, 12. In its publication *Major Uses of Land in the United States* for 1969, 1974, and 1992, the USDA estimates rates of development, but its estimate for the 1990s is just under one million acres a year. However, the USDA figure is only for land categorized as "urban."

10. Louis Uchitelle, "Can the New Economy Navigate Rougher Waters?" *New York Times*, 18 December 2000, C1.

11. William J. Baumol and Wallace E. Oates, "Conservation of Resources and the Price System," in *Economics of Resources*, ed. Robert D. Leiter and Stanley L. Friedlander (New York: Cyrco Press, 1976), 3:21.

12. Jad Mouawad, "As Geopolitics Takes Hold, Cheap Oil Recedes into the Past," *New York Times*, 3 January 2005, C4.

13. James McCusker and Russell R. Menard, *Economy of British America, 1607–1789* (Chapel Hill: University of North Carolina Press, 1995), 141. See also Jeffrey G. Williamson and Peter H. Lindert, *American Inequality* (New York: Academic Press, 1980), 10–11.

14. Robert Heilbroner, "The Human Prospect," *New York Review*, 24 January 1974, 21–34.

15. This is a paraphrase of the last line of Louise Gluck's "The Racer's Widow," *The American Poetry Anthology*, ed. Daniel Halpern (New York: Avon Books, 1975), 111.

CHAPTER 2

1. Samuel Eliot Morison, *The European Discovery of America: The Northern Voyages* (New York: Oxford University Press, 1971), 344–49, and Mark Kurlansky, *Cod: A Biography of the Fish That Changed the World* (New York: Penguin, 1997).

2. McCusker and Menard, *Economy of British America*, 98.

3. Jeffrey P. Brain, "Fort Saint George," Peabody Essex Institute, Salem, MA, 1995.

4. Morison, *Builders of the Bay Colony*, 8.

5. John Smith, *The Complete Works of Captain John Smith*, ed. Phillip Barbour (Chapel Hill: University of North Carolina Press, 1986), 332–33.

6. Morison, *Builders of the Bay Colony*, 13. Morison suggests a similarity between the colonization scheme of Rev. John White, a chief founder of the Massachusetts Bay Colony, and the ideas in Smith's pamphlet, *A Description of New England*, published in 1616 (25).

7. McCusker and Menard, *Economy of British America*, 117–18.

8. New Plymouth Colony, *Records of the Colony of New Plymouth in New England*, ed. Nathaniel Shurtleff (New York: AMS Press, 1968), 16.

9. Ibid., 14. In the first division, the original inhabitants were assigned between one and six acres according to their status, but some of this land was outside the palisade.

10. John Stilgoe, *Common Landscape of America, 1580–1845* (New Haven: Yale University Press, 1982), 216.

11. Staughton Lynd, *Class Conflict, Slavery, and the United States Constitution* (Westport, CT: Greenwood Press, 1980), 25–77.

12. New Plymouth Colony, *Records of the Colony of New Plymouth*, 210.

13. This point is made in relation to fence building in Diana Muir's *Reflections in Bullough's Pond* (Hanover, NH: University Press of New England, 2000).

14. Alice Hanson Jones, *Wealth of a Nation to Be* (New York: Columbia University Press), 225.

15. Lee Nathaniel Newcomer, *Embattled Farmers: A Massachusetts Countryside in the American Revolution* (New York: Columbia University, King's Crown Press, 1953), 15.

16. Ibid., 16.

17. Thomas Jefferson drafted the Land Ordinance of 1784, as described by Benjamin Horace Hibbard in *The History of the Public Land Policies* (Madison: University of Wisconsin Press, 1965).

18. Hibbard, *History of the Public Land Policies*, 181–83.

19. Ralph Waldo Emerson, "The Present Age," in *The Early Lectures of Ralph Waldo Emerson. Vol. III, 1838–1842*, ed. Robert E. Spiller and Wallace E. Williams (Cambridge, MA: Belknap Press, 1972), 196.

CHAPTER 3

1. Muir, *Reflections in Bullough's Pond*, 142.

2. Henry David Thoreau, *Cape Cod*, ed. Jos. J. Moldenhauer (Princeton, NJ: Princeton University Press, 1988), 27.

3. John Kenneth Galbraith, *The Great Crash 1929* (Boston: Houghton Mifflin, 1979), 7.

4. Henry C. Kittredge, *Cape Cod: Its People and Their History* (Orleans, MA: Parnassus Imprints, 1950), 306.

5. The following items are from the archives of the Mashpee Historical Society in notebooks on the development of Poponessett: "Tents and Trailers Shelter Happy Colony at Poponessett Beach," *Falmouth Enterprise*, undated; "Building at Poponesset [*sic*]," *Falmouth Enterprise*, 8 April 1938; Ad for Poponessett Beach Tourist Camp; and Letter, Popponessett Beach, Inc., 25 April 1940.

CHAPTER 4

1. Supreme Court of Virginia, *Board of County Supervisors of Fairfax County, Virginia v. G. Wallace Carper, et al.*, Record No. 4865, 200 Va. 653, 107 S.E.2d 390, 1959 Va. Lexis 151, 16 March 1959.

2. Steve Libby, Cheryl Court, and Lee Epstein, *Mapping a Future for the Washington, D.C., Metropolitan Region*, Joint study by the University of Maryland,

U.S. Geological Survey, and Chesapeake Bay Foundation. (Chesapeake Bay Foundation, 2002).

3. "Arlington Politics," *Washington Post*, 27 November 2002.

4. During this period, Congress passed the Wilderness Act of 1964; the Wild and Scenic Rivers Act of 1968; the National Environmental Policy Act of 1969, creating the Environmental Protection Agency; the Clean Air Act of 1970; the Water Pollution Control Act of 1972; the Endangered Species Acts of 1973, 1978, and 1982; the National Forest Management Act of 1976; the Federal Land Policy and Management Act of 1976; and the Alaska National Interest Lands Conservation Act of 1980. And in 1980, responding to Love Canal, the Environmental Protection Agency established its toxic cleanup Superfund.

CHAPTER 5

1. Loudoun County Department of Economic Development, *2001 Annual Growth Summary*, Loudoun County, Virginia, 7–28.

2. Loudoun County Board of Supervisors Public Hearing, 6 November 2002 (Virginia: Oak Grove Reporting), 2002.

3. The zoning enforced twenty-acre to fifty-acre lots but allowed smaller (ten-acre and twenty-acre) lots if the houses were clustered together, leaving common open areas.

4. Michael Laris, "Loudoun Faction Was Set for Fight," *Washington Post*, 13 March 2005, C1.

5. Michael Laris, "Loudoun Developer Leaves State Board," *Washington Post*, 17 March 2005, B4.

6. Will Rogers, "It's Easy Being Green," *New York Times*, 20 November 2004.

CHAPTER 6

1. George Washington, Letter to Henry Lee, In Congress, 18 June 1786. *Writings of George Washington, Vol. 11, 1785–1789*, ed. Worthington Chauncey Ford (New York: G. P. Putnam, Knickerbocker Press, 1891).

2. George Washington, Letter to Henry L. Charton, 20 May 1786, ibid.

3. Francis Jennings, *Empire of Fortune* (New York: W. W. Norton, 1988), 62n48.

4. George Washington, *The Papers of George Washington, Retirement Series. Vol. 4, April–December 1799*, ed. W. W. Abbot (Charlottesville: University of Virginia Press, 1999), 519–27n9–10.

5. Jennings, *Empire of Fortune*, 62n48.

6. George Washington, "Three Plats of George Washington's Land on the Ohio River, January 1775," George Washington Papers, 1741–1799, series 4, General Correspondence, 1697–1799 (Washington, DC: Library of Congress).

7. Robert Leslie Jones, *History of Agriculture in Ohio to 1880* (Kent, OH: Kent State University Press, 1983), 20.

8. Washington detailed his claims in his 20 May 1786 letter to Henry L. Charton. His will substantiates the claims and adds another thousand acres.

9. George Washington, "Schedule of Property, The Will of George Washington," in *Papers of George Washington, Retirement Series*, Documents and Articles, http://gwpapers.virginia.edu/documents/will/property.html.

10. George Washington's family and land history in Virginia can be traced to his great-grandfather, John Washington, a sailor whose ship went aground in the Potomac and who, like other members of the stranded crew, was awarded a grant of land from King Charles II.

11. Daniel Friedenberg, *Life, Liberty, and the Pursuit of Land* (Buffalo, NY: Prometheus Books, 1992), 36.

12. Ibid., 41–43.

13. David Herbert Donald, *Lincoln* (New York: Simon & Schuster/Touchstone, 1995), 21.

14. As quoted in Jennings, *Empire of Fortune*, 60.

15. George Washington, Letter to Thomas Jefferson, 31 August 1788.

16. As quoted in William B. Scott, *In Pursuit of Happiness: American Conceptions of Property from the Seventeenth to the Twentieth Century* (Bloomington: Indiana University Press, 1977), 41.

17. Thomas Jefferson, Letter to James Madison, 28 October 1785.

18. As quoted in Friedenberg, *Life, Liberty, and the Pursuit of Land*, 359.

19. As quoted in Lee Soltow, "Inequality in Abundance? Land Ownership in Early 19th-Century Ohio," *Ohio History* 88 (1977): 134.

20. De Toqueville, *Democracy in America*, 64.

21. Though the land could have provided even more social benefits, prejudice prevented it. Many groups, including African Americans and most Native Americans, were excluded from the government's largesse. By the Revolutionary War, only about 18 percent of women owned land. The Homestead Act of 1862 officially allowed single women, not married women, to own land. Furthermore, even when land was free, many people could not afford to outfit a farm. Still another group—the disabled—lacked the physical ability to work the farm, as did most people over the age of forty-five—the small percentage that survived to that age in colonial and early America. In another example, in 1854, Franklin Pierce vetoed a widely supported bill appropriating ten million acres to aid the "indigent insane," as described in James W. Oberly, *Sixty Million Acres: American Veterans and the Public Lands before the Civil War* (Kent, OH: Kent State University Press, 1990). Children, many of them orphaned and unprotected by child labor laws, constituted an extremely vulnerable social group.

22. Hibbard, *History of the Public Land Policies*, 34.

23. Muir, *Reflections in Bullough's Pond*, 87–89.

24. William Dollarhide, *Map Guide to American Migration Routes, 1735–1815* (Bountiful, UT: Heritage Quest), 13.

25. Estelle May Stewart, *History of Wages in the United States from Colonial Times to 1928*, Bulletin of the United States Bureau of Labor Statistics No. 604 (Washington, DC: Government Printing Office, 1934), 21.

26. Hibbard, *History of the Public Land Policies*, 33.

27. Ibid., 35.

28. Paul Schuster Taylor, *Labor on the Land: Collected Writings 1930–1970* (New York: Arno Press), 179–87.

29. Daniel Friedenberg points out, *in Life, Liberty, and the Pursuit of Land*, that the delegates to the Constitutional Convention were almost exclusively large landowners (321–23).

30. Theodore Roosevelt, *Reform of the Land Laws: Conservation of National Resources*. Extracts from *Recommendations of the President, the Secretary of the Interior, and the Commissioner of the General Land Office, etc.* (Washington: Government Printing Office, 1910), doc. 283.

CHAPTER 7

1. Soltow, "Inequality in Abundance?" 136.

2. McCusker and Menard, *Economy of British America*, 141.

3. Soltow, "Inequality in Abundance?" 136.

4. Jones, *History of Agriculture in Ohio*, 34.

5. U.S. Department of Commerce, *Historical Statistics of the United States*, 163–64.

6. John J. Horton, *The Jonathan Hale Farm* (Cleveland: Western Reserve Historical Society, 1990), 53.

7. Ibid., 52–93.

8. Comparing Hale's holdings to figures in Soltow, "Inequality in Abundance?" 143.

9. Donald, *Lincoln*, 22.

10. Ibid., 19.

11. Robert E. Gallman and John Joseph Wallis, *American Economic Growth and Standards of Living before the Civil War* (Chicago: University of Chicago Press, 1992), 13.

12. Martin Shanahan, "How Much More Unequal? Consistent Estimates of the Distribution of Wealth in the United States between 1774 and 1860." Jeffrey G. Williamson and Peter Lindert, *American Inequality: A Macroeconomic History* (New York: Academic Press, 1980), 5.

13. Gallman and Wallis, *American Economic Growth*, 13–14.

14. Gallman and Wallis, *American Economic Growth*, 8–9.

15. Wage information comes from U.S. Department of Commerce, *Historical Statistics of the United States*, 163–64, and from Stewart, *History of Wages*, esp. 138, 143–354.

16. In the words of Jeffrey G. Williamson and Peter H. Lindert, in *American Inequality:* "When land prices shot up, relative to wages, inequality ensued" (10).

17. It is hard to think of a cost of living that is not in some way tied to land values. For example, while land and housing costs rise, store rents rise, and the rise is passed to consumers in the price of goods. Similarly, professional fees reflect the needs of professionals to pay their own rents or mortgages.

18. Jones, *History of Agriculture in Ohio*, 320.

19. W. A. Lloyd, J. I. Falconer, and C. E. Thorne, *The Agriculture of Ohio. Vol. 1918–1919*, Bulletin 326 of the Ohio Agricultural Experiment Station, Wooster, OH, 1918.

20. Cindy Hahamovitch, *The Fruits of Their Labor: Atlantic Coast Farmworkers and the Making of Migrant Poverty* (Chapel Hill: University of North Carolina Press, 1997), 17.

21. Six hundred thousand citizens eventually took advantage of the Homestead Act to claim land, though an indeterminate number of them were speculators circumventing the rules.

22. *The Southern Cultivator*, 1889, 213, 484; Jones, *History of Agriculture in Ohio*, 319.

23. Jones, *History of Agriculture in Ohio*, 319.

24. *Southern Cultivator*, March 1889, 111.

25. This is not to discount the back-to-the-land movements that periodically sweep our culture or the current slight resurgence of small farms whose owners supplement farm income with income from other employment.

26. Taylor, *Labor on the Land*, 179–87.

27. Carl Taylor, Louis J. Ducoff, Margaret Jarman Hagood, *Trends in the Tenure Status of Farmworkers in the United States since 1880* (Washington, DC: United States Department of Agriculture), 11.

28. A thorough and moving account of these losses is given by Brett Williams, who chronicles the loss of small privileges such as fishing among the population of the Anacostia neighborhoods of Washington, DC, in "A River Runs Through Us," *American Anthropologist* 103, no. 2 (2001): 409–31.

CHAPTER 8

1. Janet Okoben, "Reading, Writing and Rebuilding," *Plain Dealer*, 29 August 2003, A1.

2. Loring Underwood, "The City Beautiful: The Ideal to Aim At," *American City*, 1910, 214.

3. S. B. Sutton, ed., *Civilizing American Cities: A Selection of Frederick Law Olmsted's Writings on City Landscape* (Cambridge: MIT Press, 1971), 88.

4. As quoted by Eugene Murdock in *The Buckeye Empire: An Illustrated History of Ohio Enterprise* (Northbridge, CA: Windsor, 1988), 70.

5. Ward-by-ward figures for population, homeownership, disease, and death were culled from census records and presented by Richard Klein as an appendix to his Ph.D. dissertation, "Nineteenth Century Land Use Decisions in Cleveland, Ohio: A Case Study of Neighborhood Development and Change in Ohio City" (PhD diss., University of Akron, 1983).

6. Tom L. Johnson, *My Story*, ed. Elizabeth Hauser (New York: B. S. Huebsch, 1911), xiii.

7. David D. Van Tassel and John Grabowski, eds., *Cleveland: A Tradition of Reform* (Kent, OH: Kent State University Press, 1986), 30.

8. Ibid., 30.

9. Kenneth Kusmer, *A Ghetto Takes Shape: Black Cleveland, 1870–1930* (Urbana: University of Illinois Press, 1978), 35–37.

10. Ibid., 35–37.

11. Ibid., 48.

CHAPTER 9

1. This is discussed at the end of chapter 10. Briefly, home building has begun to take away productive agricultural land, timberlands, and access to oil and minerals.

2. John Stuart Mill, *Principles of Political Economy*, ed. Donald Winch (New York: Penguin Books, 1970), 380.

3. Ibid., 384.

4. Ibid., 126.

5. Robert V. Andelson, ed., *Critics of Henry George: A Centenary Appraisal of Their Strictures on Progress and Poverty* (London: Associated University Presses, 1979), 19–26, 30.

6. Wallace E. Oates and Robert Schwab, "The Impact of Urban Land Taxation: The Pittsburgh Experience," *National Tax Journal* 50:1–21.

7. Barbara Clemenson, "The Political War against Tom L. Johnson, 1901–1909" (master's thesis, Cleveland State University, 1989), 99–101.

8. To be clear, Ohio created statutes in its original constitution limiting black Americans' civil rights, including the right to own property. But by the time of the Civil War, Cleveland had strong abolitionist sentiments, too, and a less hostile attitude toward African Americans than many other American cities, according to Kenneth Kusmer in *A Ghetto Takes Shape.*

CHAPTER 10

1. Dolores Hayden, *Building Suburbia: Green Fields and Urban Growth, 1820–2000* (New York: Pantheon Books, 2003), 89–90.

2. Texas Low Income Housing Information Service, "The Public Housing Debate," 1988, 1, http://www.texashousing.org/txlihis/phdebate/past1.html.

3. Van Tassel and Grabowski, *Cleveland: A Tradition of Reform, 16.*

4. Ibid., 30–38.

5. David D. Van Tassel and John J. Grabowski, *The Encyclopedia of Cleveland History* (Indianapolis: Indiana University Press, 1996), 828.

6. United States Department of Agriculture, *Major Uses of Land in the United States* (Washington, DC: Government Printing Office, 1969), 5.

7. Texas Low Income Housing Information Service, "The Public Housing Debate."

8. Ibid.

9. "Public Housing for Cleveland's Citizens," Kelvin Smith Library, Case Western Reserve University, 1968, 4.

10. National Association of Realtors, http://www.realtor.org/realtororg.nsf/pages/narhistory.

11. Hayden, *Building Suburbia*, 89–90, 97–98.

CHAPTER 11

1. Joint Center for Housing Studies, Harvard University, *The State of the Nation's Housing 2005* (Cambridge, MA: President and Fellows of Harvard College, 2005), 3.

2. Robert David Sullivan and Rachel Dyette Werkema, "Just How Unaffordable Is the Bay State?" *Boston Globe*, 9 May 2004, D4.

3. Joint Center for Housing Studies, Harvard University, *The State of the Nation's Housing 2004* (Cambridge, MA: President and Fellows of Harvard College, 2004), 4.

4. Jane Jacobs included this idea in her chapter "Subsidizing Dwellings" in *Death and Life of Great American Cities* (New York: Vintage Books, 1961), and Robert Fishman amplified it in his article "Rethinking Public Housing," *Places* 16, no. 2 (spring 2004).

5. Fishman, "Rethinking Public Housing," 28–29.

6. Housing starts in 1950, as reported in the NAHB report "A Century of Progress," were 1.95 million for a population of 150 million and are now roughly the same according to the 2005 NAHB report, "Housing Facts, Figures, and Trends." In 2004, according to *The State of the Nation's Housing* 2005, by the Joint Center for Housing Studies of Harvard University, all housing starts, including single-family, multifamily, and condominium units, were over 2 million for a population of about 293 million. In 1975, however, housing starts were only about 1.5 million, beginning a period when housing prices rose notably faster than income.

7. Joint Center for Housing Studies, Harvard University, *The State of the Nation's Housing 2003* (Cambridge, MA: President and Fellows of Harvard College, 2003).

8. As quoted in Scott Greenberger, "Report Rates Boston Most Expensive City," *Boston Globe*, 8 September 2005.

9. Articles on timberlands lost in Maine and Washington State appeared in the *Boston Globe* ("A Race to Save the Maine Woods," 18 September 2000, and "Houses, Resorts Planned for North Woods," 5 April 2005) and *Growth/No Growth* ("Model Land Trust Abandoned," April 2003, summarizing an 11 March 2003 article in the *Seattle Times*, "Preservation Deal for Forest Falls Through"). An article on home building blocking access to oil under Los Angeles appeared in the *New York Times*, 8 December 2002, and was summarized in *Growth/No Growth*, January 2003, 9. An article on productive Idaho farmland giving way to subdivisions, "Idaho: Loss of Farmland Hurts Local Economies," appeared in *Growth/No Growth*, September 2002. An article on Nantucket bed and breakfasts being bought for homes, "As Profits Slip, Inns on Nantucket Are Converted to Homes," appeared in the *Boston Globe*, 24 May 2004. An article on the same phenomenon on Cape Cod, "What's Killing the B&B's of Cape Cod?" appeared in the *New York Times*, 19 August 2005.

CHAPTER 12

1. Oberly, *Sixty Million Acres*, 130.

2. Liberal grazing laws led to widespread destruction of the public grazing lands, resulting in the Taylor Grazing Act of 1934. The relationship between timber policies and environmental protections is still debated today. Ted Steinberg's discussion of the rise of the timber industry in *Down to Earth: Nature's Role in American History* (New York: Oxford University Press, 2002), 62–67, is roughly summarized by this excerpt: "The government subsidized large-scale timber production mainly through the sale, at bargain prices, of federal timberland. By allowing lumber companies to amass huge quantities of inexpensive land, it underwrote the industry's growth" (65).

3. Oberly, *Sixty Million Acres*, 130, 161. According to Oberly, very little public land was placed on the market during the Pierce administration of 1856, while during the previous administration all of Iowa was sold.

4. Reuel Shinnar discusses the greater energy and labor requirements for mining oil from shale in "Net Energy or Energy Analysis," in *The Economics of Resources*, ed. Leiter and Friedlander, 55. He and other contributors to *The Economics of Resources* discuss other phenomena affecting the prices of resources as they become scarcer, which include rising prices as the resource becomes rarer and more difficult to recover and prices that may fall briefly as demand slackens in response to the higher prices.

5. "The Importance of Being Private," *Forbes*, 29 November 2004, 201; "Private Mission," *Forbes*, 29 November 2004, 204.

6. Ted Steinberg, *Down to Earth* (New York: Oxford University Press, 2002), 116.

7. Douglas Jehl, "Gold Miners Eager for Bush to Roll Back Clinton Rules," *New York Times*, 16 August 2001.

8. Russell Elliott, *History of Nevada* (Lincoln: University of Nebraska Press, 1973), 72–73.

9. Tasker Oddie, letters of 16 February 1898, 15 August 1899, in *Letters from the Nevada Frontier*, ed. William A. Douglas and Robert Nylen (Reno: University of Nevada Press, 1992), 5, 6, 129.

10. Ibid., 211–14.

11. Ibid., 125.

12. Ibid., 136.

13. Letter from Oddie to his mother, 4 September 1900, in *Letters from the Nevada Frontier*, ed. Douglas and Nylen, 225.

14. Douglas and Nylen, *Letters from the Nevada Frontier*, xxi–xxii.

15. Letter from Oddie to his mother, 27 August 1899, in *Letters from the Nevada Frontier*, ed. Douglas and Nylen, 134.

16. Letter from Oddie to his mother, 7 August 1900, in *Letters from the Nevada Frontier*, ed. Douglas and Nylen, 214.

CHAPTER 13

1. Frederick Jackson Turner, *The Frontier in American History* (New York: Henry Holt, 1920), 129.

2. Erling Nicolai Rolfsrud, *The Story of North Dakota* (Alexandria, MN: Lantern Books, 1963), 1.

3. As quoted by Rolfsrud, *Story of North Dakota*, 1.

4. Colonel William R. Marshall, "Journal of Military Expedition Against Sioux, 1863, Under Command of Brig. General Henry Hastings Sibley, 7th Minnesota," Special Collections, Chester Fritz Library, University of North Dakota.

5. Hiram M. Drache, *The Day of the Bonanza: A History of Bonanza Farming in the Red River Valley of the North* (Fargo: North Dakota Institute for Regional Studies, 1964), 4.

6. McCusker and Menard, *The Economy of British North America.*

7. *U.S. Census 1910*, Geostat Center of the Fisher Library of the University of Virginia, http://fisher/lib.virginia.edu/collections/stats/histcensus/.

8. The map accompanies *Blanchard's Citizen's Guide to Chicago*, Pusey Map Library, Harvard University.

9. Hayden, *Building Suburbia*, 98.

10. United States Department of Commerce, *Statistical Abstract of the United States 2003* (Washington, DC: Government Printing Office, 2003), table 740.

11. Calculated on the basis of figures shown in table 740, "Corporations by Receipt-Size Class and Industry, 1999 and 2000," in U.S. Department of Commerce, *Statistical Abstract of the United States 2003*.

12. Ibid., table 741, "Corporations by Asset-Size Class and Industry: 2000."

CHAPTER 14

1. California Department of Food and Agriculture, *California Agricultural Statistics, 2003*, www.cdfa.ca.gov.

2. These were found in Imperial Valley.

3. Western Management Consultants, *Water and San Diego County Growth: A Study for the San Diego County Water Authority* (Phoenix: Western Management Consultants, 1966), 147–48.

4. James S. Brown, *Life of a Pioneer: Being the Autobiography of James S. Brown* (Salt Lake City: G. Q. Cannon, 1900), 273.

5. Western Management Consultants, *Water and San Diego County Growth.*

6. Paul Schuster Taylor, "The Excess Land Law: The Execution of a Public Policy," in *Essays on Land, Water, and the Law in California* (New York: Arno Press, 1979), 478.

7. Different sources give different statistics. For example, F. C. Farr, in *The History of Imperial County, California* (Berkeley, CA: Elms and Franks, 1918), 16, mentions nine thousand people growing 150,000 acres of crops. The article "A History of the Imperial Valley," Part III, by the Imperial County Historical Society, gives a population of twelve thousand for 1905, http://www.imperial.cc.ca.us/Pioneers/HISTORY4.HTM.

8. Farr, *History of Imperial County*, 63.

9. U.S. Department of Commerce, *Historical Statistics of the United States.*

10. Bettina Boxall, "Water Pacts Give State's Growers New Profit Stream," *Los Angeles Times*, 16 February 2005.

11. Dean E. Murphy, "Pact in West Will Send Farms' Water to Cities," *New York Times*, 17 October 2003, A1.

12. Boxall, "Water Pacts."

13. Janet Wilson, "EPA Fights Waste Site Near River," *Los Angeles Times*, 5 March 2005.

14. Winnie Hu, "To Protect Water Supply, City Acts as a Land Baron," *New York Times*, 9 August 2004, A14.

15. Timothy Egan, "Paying on the Highway to Get Out of First Gear," *New York Times*, 28 April 2005, A1. Average time spent in traffic is listed by city in the Texas Transportation Institute's *2005 Urban Mobility Report* (College Station: Texas A&M University).

CHAPTER 15

1. Rogers, "It's Easy Being Green," A31.

2. Kirk Johnson, "Counting the Costs of Growth with a Forest of Formulas," *New York Times*, 23 November 2003, 12wk.

3. This 1991 study by Hartig, Mang, and Evans ("Restorative Effects of Natural Experiences," *Environment and Behavior* 23, no. 3, 26) was summarized in Nancy M. Wells, "At Home with Nature," *Environment and Behavior* 32, no. 6 (November 2000): 780.

4. Nancy M. Wells, "Nearby Nature: A Buffer of Life Stress among Rural Children," *Environmental Behavior and Psychology* (35, no. 3): 312.

5. Wells, "At Home with Nature," and "Nearby Nature."

6. Marjorie Kinnan Rawlings, *The Yearling* (New York: Charles Scribner's Sons, 1939), 18.

7. Don P. Blayney and Alden C. Manchester, "Large Companies Active in Changing Dairy Industry," *FoodReview* 23, no. 2 (2000): 1–2.

8. David Barboza, "Wisconsin's Cheese State Fights to Stay That Way," *New York Times*, 28 June 2001, C1.

9. Eric Schlosser, *Fast Food Nation* (Boston: Houghton Mifflin, 2001), 5.

10. Carla Occaso, "Farmer Sends Manure Instead of Milk," *Caledonian-Record News*, 23 October 2005, http://www.newfarm.org/news/2002/111402/manure_milk.shtml.

11. Dean Foods, http://www.deanfoods.com/aboutus/history.asp.

12. Schlosser, *Fast Food Nation*, 6.

13. Gretchen Morgenson, "Explaining (Or Not) Why the Boss Is Paid So Much," *New York Times*, 25 January 2004, C1.

14. According to a Boston Consulting Group study quoted in Ross Douthat, "Primary Sources: Spoil the CEO . . . ," *Atlantic Monthly*, May 2005, 52, both outsize expectations for corporate performance and outsize incentives for executives to achieve them were found in the twenty-five companies perpetrating the largest cor-

porate frauds. Those twenty-five CEOs received "eight times as much stock-based pay" as CEOs employed by their more honest competitors.

15. Frank Levy, *The New Dollars and Dreams: American Incomes and Economic Change* (New York: Russell Sage Foundation, 1998).

16. Steven Greenhouse, "Report Shows Americans Have More 'Labor Days,'" *New York Times*, 1 September 2001.

17. Bill Carter and Jim Rutenberg, "Creative Voices Say Television Will Suffer in New Climate," *New York Times*, 3 June 2003, C1.

18. Three of these suits are described in Jennifer Lee, "7 States to Sue E.P.A. Over Standards on Air Pollution," *New York Times*, 21 February 2003.

19. Congress authorized funding annually through the Department of Energy for the Hybrid and Electric Vehicle Program. The program's annual reports are available in the Government Documents section of many libraries.

20. U.S. Department of Energy, *Electric and Hybrid Vehicles Program: 18th Annual Report to Congress for Fiscal Year 1994* (Washington, DC: Government Printing Office, 1995), 1–1.

21. The toxic gases released by auto emissions have been the focus of reforms for a couple of decades. The current threat comes mainly from fine particles, as explained in an article by Jonathan Shaw based on research at the Harvard School of Public Health, "Clearing the Air," *Harvard Magazine*, May–June 2005. Power plants also emit fine particles, as do household activities such as cooking and grilling. The life expectancy of Americans living in the six cities with the dirtiest air is two years shorter than that of the rest of the population.

BIBLIOGRAPHY

SECTION 1
PERPETUAL GROWTH IN THE LAND OF ABUDANCE

Arnold, David. "U.S. versus the World: American Consumerism Is Still the Focus of a Global Environmental Debate." *Boston Globe*, 20 August 2002, D1.

Becker, Elizabeth. "2 Acres of Farm Lost to Sprawl Each Minute, New Study Says." *New York Times*, 4 October 2002, A19.

Bradley, Ian. "The Rise of the Middle Classes." In *The English Middle Classes*. London: Wm. Collins Sons, 1982.

Energy Information Administration. *The Changing Structure of the Electric Power Industry*. Washington, DC: Government Printing Office, 1993.

———. *The Changing Structure of the Electric Power Industry: An Update*. Washington, DC: Government Printing Office, 1996.

Galbraith, John Kenneth. *The Affluent Society*. Boston: Houghton Mifflin, 1998.

Harris, Martin. *Cannibals and Kings: The Origins of Cultures*. New York: Random House, 1977.

Heilbroner, Robert. "The Human Prospect." *New York Review*, 24 January 1974.

Johnson, Barry. "Personal Wealth, 1992–1995." *Statistics of Income Bulletin*, winter 1998. http://www.irs.gov/taxstats/.

Jones, Alice Hanson. *Wealth of a Nation to Be*. New York: Columbia University Press, 1978.

King, Roger, and John Raynor. *The Middle Class*. 2d ed. London: Longman Group, 1981.

Krugman, Paul. "For Richer: How the Permissive Capitalism of the Boom Destroyed American Equality." *The New York Times Sunday Magazine*, 20 October 2002, 62.

Leiter, Robert D., and Stanley L. Friedlander, eds. *The Economics of Resources*. New York: Cyrco Press, 1976.

Levy, Frank. *The New Dollars and Dreams: American Incomes and Economic Change*. New York: Russell Sage Foundation, 1998.

Macaulay, Thomas Babington. *The Selected Letters of Thomas Babington Macaulay*. Ed. Thomas Pinney. Cambridge: Cambridge University Press, 1982.

Technology Futures, Inc., and Scientific Foresight, Inc. *Principles for Electric Power Policy*. Westport, CT: Quorum Books, 1984.
Texas Transportation Institute. *2001 Urban Mobility Report*. College Station: Texas A&M University, 2001, 2005.
Tocqueville, Alexis de. *Democracy in America*. Ed. J. P. Mayer. 12th ed. Garden City, NY: Doubleday/Anchor Books, 1969.
United States Censuses 1790–1960. Geostat Center of the Fisher Library of the University of Virginia. http://fisher/lib.virginia.edu/collections/stats/histcensus/.
United States Census Bureau. "The Changing Shape of the Nation's Income Distribution 1947–1998." U.S. Department of Commerce, Economics and Statistics Administration, 2000.
United States Department of Agriculture. *Major Uses of Land in the United States*. Washington, DC: Government Printing Office, 1969, 1974, 1992.
United States Department of Agriculture and United States Bureau of the Census. *Graphic Summary of Land Utilization in the U.S.* Washington, DC: Government Printing Office, 1945.
United States Department of Commerce. *Historical Statistics of the United States: Colonial Times to 1970*. Part I. Washington, DC: Government Printing Office, 1975.

SECTION 2
LAND AS PROPERTY: COLONIAL AMERICA: CARVING UP HEAVEN

An Act Revising the Municipal Zoning Laws. General Court of Massachusetts, General Laws, Chapter 269. Boston: General Court, 1933.
Bradford, William. *Bradford's History of Plimoth Plantation from the Original Manuscript*. Boston: Wright & Potter, 1899.
Brain, Jeffrey. "Fort Saint George." Peabody Essex Institute, Salem, MA, 1995.
Demos, Jonathan. *Entertaining Satan: Witchcraft and the Culture of Early New England*. New York: Oxford University Press, 1982.
———. "Images of the Family, Then and Now." In *Changing Images of the Family*, ed. Virginia Tufte, 43–60. New Haven: Yale University Press, 1979.
———. *A Little Commonwealth*. New York: Oxford University Press, 1970.
Denman, D. R. *Origins of Ownership*. London: George Allen & Unwin, 1958.
Friedenberg, Daniel M. *Life, Liberty, and the Pursuit of Land*. Buffalo, NY: Prometheus Books, 1992.
Galbraith, John Kenneth. *The Great Crash 1929*. Boston: Houghton Mifflin, 1979.
Gallatin, Albert. *Report of the Secretary of the Treasury, on the subject of public roads and canals; made in pursuance of a resolution of Senate, of March 2, 1807*. Washington, DC: The Senate, 1808.
Greenhouse, Linda. "Justices Appear Reluctant to Increase Land-Use Oversight." *New York Times*, 23 February 2005, A12.
———. "Justices Weaken Movement Backing Property Rights." *New York Times*, 24 April 2002, A1.

Greven, Phillip. *Four Generations: Population, Land, and Family in Colonial Andover, Massachusetts.* Ithaca, NY: Cornell University Press, 1970.

Hibbard, Benjamin Horace. *The History of the Public Land Policies.* Madison: University of Wisconsin Press, 1965.

Jacobs, Harvey M. "The Future of an American Ideal." In *Private Property in the 21st Century: The Future of an American Ideal,* ed. Harvey M. Jacobs. In Association with the Lincoln Institute of Land Policy. Northampton, MA: Edward Elgar, 2004.

———. "Introduction." In *Private Property in the 21st Century: The Future of an American Ideal,* ed. Harvey M. Jacobs. In Association with the Lincoln Institute of Land Policy. Northampton, MA: Edward Elgar, 2004.

Jefferson, Thomas. *Writings.* Ed. Merrill D. Peterson. New York: Literary Classics of the United States, 1984.

Jennings, Francis. *Empire of Fortune: Crowns, Colonies & Tribes in the Seven Years War in America.* New York: W. W. Norton, 1988.

Jones, Alice Hanson. *Wealth of a Nation to Be.* New York: Columbia University Press, 1978.

Joyce, Amy. "Cultivating New Territory." *Washington Post,* 6 December 2003, E1.

Kantor, Shawn Everett. *Politics and Property Rights: The Closing of the Open Range in the Postbellum South.* Chicago: University of Chicago Press, 1998.

Kayden, Jerold S. "Celebrating Penn Central: How the Supreme Court's Preservation of Grand Central Terminal Helped Preserve Planning Nationwide." *Planning,* June 2003, 10.

———. "Charting the Constitutional Course of the Private Property: Learning from the 20th Century." In *Private Property in the 21st Century: The Future of an American Ideal,* ed. Harvey M. Jacobs. In Association with the Lincoln Institute of Land Policy. Northampton, MA: Edward Elgar, 2004.

Kittredge, Henry C. *Cape Cod: Its People and Their History.* Orleans, MA: Parnassus Imprints, 1950.

Kurlansky, Mark. *Cod: A Biography of the Fish That Changed the World.* New York: Penguin, 1997.

Laris, Michael. "Loudoun Developer Leaves State Board." *Washington Post,* 13 March 2005.

———. "Loudoun Faction Was Set for Fight." *Washington Post,* 13 March 2005, C1.

Libby, Steve, Cheryl Court, and Lee Epstein. *Mapping a Future for the Washington, D.C., Metropolitan Region.* Joint study by the University of Maryland, U.S. Geological Survey, and Chesapeake Bay Foundation. Annapolis: Chesapeake Bay Foundation, 2002.

Loudoun County Board of Supervisors Public Hearing. 2 November 2002. Virginia: Oak Grove Reporting, 2002.

———. 6 November 2002. Virginia: Oak Grove Reporting, 2002.

Loudoun County Department of Economic Development. *2001 Annual Growth Summary.* Loudoun County, VA, 2001.

Loudoun County Planning Department. *Proposed Revisions to the 1993 Zoning Ordi-*

nance, ZOAM 2002–0003, ZMAP 2002–0014, For Public Hearing November 2, 2002 and November 6, 2002. Loudoun County, Virginia, 2002.

McCusker, James, and Russell R. Menard. *The Economy of British America, 1607–1789.* Chapel Hill: University of North Carolina Press, 1991.

Morison, Samuel Eliot. *Builders of the Bay Colony.* Boston: Houghton Mifflin, 1930.

———. *The European Discovery of America: The Northern Voyages.* New York: Oxford University Press, 1971.

Newcomer, Lee Nathaniel. *The Embattled Farmers: A Massachusetts Countryside in the American Revolution.* New York: Columbia University, King's Crown Press, 1953.

New Plymouth Colony. *Records of the Colony of New Plymouth in New England.* Ed. Nathaniel Shurtleff. Boston: AMS Press, 1968.

Nineteenth Judicial Circuit of Virginia. *Aldre Properties Inc. v. Board of Supervisors of Fairfax County, Virginia, et al.* Chancery nos. 78463-A, 78476, 78450, 78425, County of Fairfax, Virginia, 7 January 1985.

Perkins, Edwin. *The Economy of Colonial America.* New York: Columbia University Press, 1980.

Price, Edward T. *Dividing the Land: Early American Beginnings of Our Private Property Mosaic.* Chicago: University of Chicago Press, 1995.

Savage, Charlie. "Limit Urged on Eminent Domain." *Boston Globe,* 23 February 2005.

Scott, William B. *In Pursuit of Happiness: American Conceptions of Property from the Seventeenth to the Twentieth Century.* Bloomington: Indiana University Press, 1977.

Smith, John. *Captain John Smith: A Select Edition of His Writings.* Ed. Karen Ordahl Kupperman. Chapel Hill: University of North Carolina Press, 1988.

Stilgoe, John. *Common Landscape of America, 1580–1845.* New Haven: Yale University Press, 1982.

Supreme Court of the United States. *Tahoe-Sierra Preservation Council, Inc., et al., v. Tahoe Regional Planning Agency et al., Certiorari to the United States Court of Appeals for the Ninth Circuit.* 535 U.S. (2002).

———. *Village of Euclid, Ohio v. Ambler Realty Co.* 272, U.S. 52 (1926).

Supreme Court of Virginia. *Board of County Supervisors of Fairfax County, Virginia v. G. Wallace Carper, et al.* Record No. 4865, 200 Va. 653; 107 S.E.2d 390; 1959 Va. Lexis 151, 16 March 1959.

———. *Board of Supervisors of Fairfax County v. Virgil Jackson.* Record No. 81241, 28 August 1980.

Thoreau, Henry David. *Cape Cod.* Ed. Jos. J. Moldenhauer. Princeton, NJ: Princeton University Press, 1988.

United States Department of Commerce. Advisory Committee on City Planning and Zoning. *A Standard State Zoning Enabling Act under Which Municipalities Can Adopt Zoning Regulations.* Preliminary Edition. Washington, DC: Government Printing Office, August 1922.

———. *A Standard State Zoning Enabling Act under Which Municipalities May Adopt Zoning Regulations.* Washington, DC: Government Printing Office, 1924, 1926.

United States Department of Commerce. "Increased Interest in Zoning Laws." Press Release, 24 August 1923. Loeb Library, Harvard University.

Washington, George. "Schedule of Property, The Will of George Washington." In *The Papers of George Washington, Retirement Series,* Documents and Articles, University of Virginia. http.//gwpapers.virginia.edu/documents/will/property.html.

SECTION 3
LAND AND WELL-BEING: EARLY NATIONHOOD: FROM FAMILY FARM TO CITY

Andelson, Robert V., ed. *Critics of Henry George: A Centenary Appraisal of Their Strictures on Progress and Poverty.* London: Associated University Presses, 1979.

Blanchard's Citizen's Guide to Chicago. Pusey Map Library, Harvard University.

Burnham, Daniel. "Report on the Group Plan of the Public Buildings of the City of Cleveland Ohio, Second Edition with Supplement Indicating the Progress of the Improvements." Western Reserve Historical Society Archives, Cleveland, OH.

Burnham, Daniel, with Group Plan Commission Members. "The Group Plan." 1903. Western Reserve Historical Society Archives, Cleveland, OH.

Cayton, Andrew R. L. *The Frontier Republic: Ideology and Politics in the Ohio Country, 1780–1825.*

Clemenson, Barbara. "The Political War against Tom L. Johnson 1901–1909." Master's thesis, Cleveland State University, 1989.

Cowan, Michael H. *City of the West: Emerson, America, and Urban Metaphor.* New Haven: Yale University Press, 1967.

Dollarhide, William. *Map Guide to American Migration Routes, 1735–1815.* Bountiful, UT: Heritage Quest, 1997.

Donald, David Herbert. *Lincoln.* New York: Simon & Schuster/Touchstone, 1995.

Emerson, Ralph Waldo. *The Early Lectures of Ralph Waldo Emerson. Vol. III, 1838–1842.* Ed. Robert E. Spiller and Wallace E. Williams. Cambridge, MA: Belknap Press of Harvard University, 1972.

———. *Emerson's Complete Works. Vol. VIII, Letters and Social Aims.* Cambridge, MA: Riverside Press, 1883.

Epstein, Helen. "Enough to Make You Sick?" *New York Times Magazine,* 12 October 2003, 75.

Firestone, David. "100,000 Could Lose Housing Subsidies, Advocates Warn." *New York Times,* 5 September 2003, A15.

Fishburn, Robert. "Rethinking Public Housing." *Places* 16, no. 2 (spring 2004)

Gallman, Robert E., and John Joseph Wallis. *American Economic Growth and Standards of Living before the Civil War.* Chicago: University of Chicago Press, 1992.

George, Henry. *Progress and Poverty.* New York: Robert Schalkenbach Foundation, 1981.

Havighurst, Walter. *Ohio: A History.* Chicago: University of Illinois Press, 2001.

Hayden, Delores. *Building Suburbia: Green Fields and Urban Growth, 1820–2000.* New York: Pantheon Books, 2003.

Hibbard, Benjamin Horace. *The History of the Public Land Policies.* Madison: University of Wisconsin Press, 1965.

Hopkins, G. M. *City Atlas of Cleveland Ohio from official records, private plans, and actual surveys.* Philadelphia: G. M. Hopkins, 1881. Map Collection, Cleveland Public Library.

———. *Plat Book of the City of Cleveland, Ohio, Vol. 1.* Philadelphia, 1932. Reserve Historical Society Archives, Cleveland, OH.

Horton, John J. *The Jonathan Hale Farm: A Chronicle of the Cuyahoga Valley.* Cleveland, OH: Western Reserve Historical Society, 1990.

Institute for the Study of Homelessness and Poverty. "Housing and Poverty in Los Angeles." July 21. www.weingart.org/institute/.

Jackson, Ronald Vern, ed. *Ohio 1850 Mortality Schedule.* Bountiful, UT: Accelerated Indexing Systems, 1979.

Johannesen, Eric. *Cleveland Architecture, 1876–1976.* Cleveland, OH: Western Reserve Historical Society, 1979.

Johnson, Barry. "Personal Wealth, 1992–1995." *Statistics of Income Bulletin,* winter 1998. http://www.irs.gov/taxstats/.

Johnson, Kirk. "Drought Settles In, Lake Shrinks and West's Worries Grow." *New York Times,* 2 May 2004, A1.

———. "Saving the Farm on Long Island Sound." *New York Times,* 13 August 2003, A20.

Johnson, Tom L. *My Story.* Ed. Elizabeth Hauser. New York: B. S. Huebsch, 1911.

———. "Statement of Mayor Tom L. Johnson on Municipal Ownership." Western Reserve Historical Society Archives, Cleveland, OH.

———. *Tom L. Johnson Papers.* Roll 1, Mss. No. 3651. Western Reserve Historical Society Archives, Cleveland OH.

———. "Tom L. Johnson's Past Utterances on Present Issues: Three-cent Fares and Other Municipal Questions." 1901. Western Reserve Historical Society Archives, Cleveland, OH.

Joint Center for Housing Studies. Harvard University. *The State of the Nation's Housing 2001.* Cambridge, MA: President and Fellows of Harvard College, 2001.

———. *The State of the Nation's Housing 2002.* Cambridge, MA: President and Fellows of Harvard College, 2002.

———. *The State of the Nation's Housing 2003.* Cambridge, MA: President and Fellows of Harvard College, 2003.

———. *The State of the Nation's Housing 2004.* Cambridge, MA: President and Fellows of Harvard College, 2004.

———. *The State of the Nation's Housing 2005.* Cambridge, MA: President and Fellows of Harvard College, 2005.

Jones, Alice Hanson. *Wealth of a Nation to Be.* New York: Columbia University Press, 1978.

Jones, Robert Leslie. *History of Agriculture in Ohio to 1880,* Kent, OH: Kent State University Press, 1983.

Kalis, Lisa. "What's Killing the B&B's of Cape Cod." *New York Times,* 19 August 2005.

Kennedy, Roger G. *Mr. Jefferson's Lost Cause: Land, Farmers, Slavery, and the Louisiana Purchase.* New York: Oxford University Press, 2003.

Keyton, Andrew R. L. *The Frontier Republic: Ideology and Politics in the Ohio Country, 1780–1825.* Kent, OH: Kent State University Press, 1986.

Klemencic, Matjaz. *Slovenes of Cleveland: The Creation of a New Nation and a New World Community.* Dolenjska, Zalozba: Novo Mesto, 1995.

Kusmer, Kenneth. *A Ghetto Takes Shape: Black Cleveland, 1870–1930*. Urbana: University of Illinois Press, 1978.

Lake, D. J. *Atlas of Cuyahoga County Ohio From actual Surveys by and under the directions of D.J. Lake, C.E.* Philadelphia: Titus, Simmons & Titus, 1874. Map Collection, Cleveland Public Library.

Lincoln, Abraham. *Speeches and Writings, 1832–1858*. Ed. Don E. Fehrenbacher. New York: Literary Classics of America, 1989.

List of Land Companies, 1600–1775. American Antiquarian Society, Worcester, MA, 2001.

Lloyd, W. A., J. I. Falconer, and C. E. Thorne. *The Agriculture of Ohio. Vol. 1918–1919*. Bulletin 326 of the Ohio Agricultural Experiment Station. Wooster, OH, 1918.

Lynd, Staughton. *Class Conflict, Slavery, and the United States Constitution*. Westport, CT: Greenwood Press, 1980.

MacKaye, Benton. "Employment and Natural Resources." Washington, DC: Government Printing Office, 1919.

Mill, John Stuart. *Principles of Political Economy*. Ed. Donald Winch. New York: Penguin Books, 1970.

Moley, Raymond. *The American Century of John C. Lincoln*. New York: Hawthorne Books, 1975.

Morris, Richard B. "Federal Land Legislation and Policy since 1789." In *Encyclopedia of American History*, 461–67. New York: Harper & Row, 1965.

Murdock, Eugene. *Buckeye Empire: An Illustrated History of Ohio Enterprise*. Northbridge, CA: Windsor Publication, 1988.

———. *Tom Johnson of Cleveland*. Dayton, OH: Wright State University Press, 1994.

National Low Income Housing Coalition. *Out of Reach: America's Housing Wage Climbs*. Washington, DC: National Low Income Housing Coalition and the Housing Assistance Council, 2003.

Navin, Rev. R. B., Wm. D. Peattie, and R. F. Stewart. "An Analysis of a Slum Area in Cleveland." Prepared for the Cleveland Metropolitan Housing Authority. Cleveland, 1933.

Oates, Wallace E., and Robert Schwab. "The Impact of Urban Land Taxation: The Pittsburgh Experience." *National Tax Journal* 50 (March 1997): 1–21.

Okoben, Janet. "Reading, Writing, and Rebuilding." *Plain Dealer*, 29 August 2003, A1.

Plunkett, Jack W. *Plunkett's Real Estate & Construction Industry Almanac*. Houston: Plunkett Research, 2003.

Porter, George H. *Ohio Politics during the Civil War Period*. New York: AMS Press, 1968.

"Public Housing for Cleveland's Citizens." Kelvin Smith Library, Case Western Reserve University, 1967 or 1968, 4.

Roosevelt, Theodore. *Reform of the Land Laws: Conservation of National Resources*. Extracts from *Recommendations of the President, the Secretary of the Interior, and the Commissioner of the General Land Office, etc.* Washington: Government Printing Office, 1910 (doc. 283).

Ross, Jaimie A., ed. "The NIMBY Report: Smart Growth & Affordable Housing." Washington, DC: National Low Income Housing Coalition, spring 2001.

Sanborn Map of Cleveland, Ohio, Vols. I, II, III, and IV, 1895 revised to 1910. Map Collection, Cleveland Public Library.

Scheiber, Harry N. *Ohio Canal Era: A Case Study of Government and the Economy, 1820–1861.* Athens: Ohio University Press, 1968.

Shanahan, Martin. "How Much More Unequal? Consistent Estimates of the Distribution of Wealth in the United States Between 1774 and 1860." *Journal of Income Distribution* 9, no. 1.

———. "In Search of Kuznet's Curve: A Reexamination of the Distribution of Wealth in the U.S. 1650–1950." In *Trends in Income Inequality during Industrialization: Proceedings of the Twelfth International Economic History Congress*, ed. L. Borodkin and P. Lindert, 39–50. Madrid, 1998.

Soltow, Lee. "Inequality in Abundance? Land Ownership in Early 19th-Century Ohio." *Ohio History* 88 (1979): 133–51.

Steckel, R. "Census Manuscript Schedules Matched with Property Tax Lists: A Source of Information on Long-Term Trends in Wealth Inequality." *Historical Methods* 27, no. 2 (spring): 71–85.

Steinberg, Ted. *Down to Earth: Nature's Role in American History.* New York: Oxford University Press, 2002.

Stewart, Estelle May. *History of Wages in the United States from Colonial Times to 1928.* Bulletin of the United States Bureau of Labor Statistics No. 604. Washington, DC: Government Printing Office, 1934.

Stilgoe, John. *Borderland: Origins of the American Suburb, 1820–1939.* New Haven: Yale University Press, 1988.

Sullivan, Robert David, and Rachel Dyette Wekema. "Just How Unaffordable Is the Bay State?" *Boston Globe*, 9 May 2004, D4.

Sutton, S. B., ed. *Civilizing American Cities: A Selection of Frederick Law Olmsted's Writings on City Landscape.* Cambridge, MA: MIT Press, 1971.

Taylor, Carl C., Louis J. Ducoff, and Margaret Jarman Hagood. *Trends in the Tenure Status of Farm Workers in the United States since 1880.* Washington, DC: Bureau of Agricultural Economics, United States Department of Agriculture, July 1948.

Taylor, Paul Schuster. *Labor on the Land: Collected Writings, 1930–1970.* New York: Arno Press, 1981.

Texas Low Income Housing Information Service. "The Public Housing Debate." 1988. http://www.texashousing.org/txlihis/phdebate/past1.html.

Underwood, Loring. "The City Beautiful: The Ideal to Aim At." *American City*, 1910, 214.

United States Census Bureau. "Geographic Comparison Table," GCT-PH1(one)-R. Population, Housing Units, Area, and Density (geographies ranked by total population): 2000, from Data Set: Census 2000 Summary File 1(one) (SF 1(one)) 100-Percent Data. http://factfinder.census.gov/.

———. "Historical Income Tables-Households." http://www.census.gov/hhes/income/histinc/h04.html.

———. Tables of Housing and Population Characteristics for American Cities, including "Quick Tables" DP-3, Profile of Selected Economic Characteristics: 2000; and DP-4, Profile of Selected Housing Characteristics: 2000. http://www.cenus.gov.

United States Department of Agriculture. "Graphic Summary of Farm Tenure in the U.S. 1948." Washington, DC.

———. "Trends in the Tenure Status of Farm Workers in the U.S. since 1880." 1948.

United States Department of Commerce. *Statistical Abstract of the United States 2003*. Washington, DC: Government Printing Office.

Van Tassel, David D., and John Grabowski, eds. *Cleveland: A Tradition of Reform.* Kent, OH: Kent State University Press, 1986.

Washington, George. *The Papers of George Washington, Retirement Series. Vol. 4, April–December 1799*. Ed. W. W. Abbot (Charlottesville: University Press of Virginia, 1999), 519–27.

———. "Schedule of Property, The Will of George Washington." In *The Papers of George Washington, Retirement Series,* Documents and Articles, University of Virginia. http.//gwpapers.virginia.edu/documents/will/property.html.

———. "Three Plats of George Washington's Land on the Ohio River, January 1775." George Washington Papers, 1741–1799, Series 4. General Correspondence, 1697–1799. Washington, DC: Library of Congress.

———. *Writings of George Washington, Vol. 11, 1785–1789, and Vol. 12, 1790–1794.* Ed. Worthington Chauncey Ford. New York: G. P. Putnam, Knickerbocker Press, 1891.

Williams, Brett. "A River Runs Through Us." *American Anthropologist* 103, no. 2 (2001): 409–31.

Williamson, J. G., and P. Lindert. *American Inequality: A Macroeconomic History.* New York: Academic Press, 1980.

SECTION 4
LAND AND ENTERPRISE: THE LATE FRONTIER: THE RACE FOR RESOURCES

Ambrose, Stephen E. *Nothing Like It in the World: The Men Who Built the Transcontinental Railroad, 1863–1869*. New York: Simon and Schuster, 2000.

Bancroft, Hubert Howe. *The Works of Hubert Howe Bancroft, Vol. XXV. History of Nevada, Colorado, and Wyoming, 1540–1888.* San Francisco: History Co., 1890.

Barringer, Felicity. "Growth Stirs a Battle to Draw More Water from the Great Lakes." *New York Times,* 12 August 2005.

Bostwick, Charles F. "Urban Sprawl Capping Oil Fields in California." *New York Times,* 8 December 2002.

Boxall, Bettina. "Water Pacts Give State's Growers New Profit Stream." *Los Angeles Times,* 16 February 2005.

Brown, James S. *Life of a Pioneer: Being the Autobiography of James S. Brown.* Salt Lake City: G. Q. Cannon, 1900.

Buehrer, Judi. "Azurix Down the Drain or Solvent Asset?" *E-Mainstream* 46, no. 2 (March–April 2002).

California Department of Food and Agriculture. *California Agricultural Statistics, 2003*. www.cdfa.ca.gov.

Calof, Rachel. *Rachel Calof's Story: Jewish Homesteader on the Northern Plains*. Ed. J. Sanford Rickoon. Bloomington: Indiana University Press, 1995.

Chapman, Sherl L. "Irrigated Agriculture, Idaho's Economic Lifeblood." Idaho Water Users Association. www.iwua.org/history/.

Congress of the United States. Congressional Budget Office. *Should the Federal Government Sell Electricity?* Washington, DC: Government Printing Office, November 1997.

Cranham, Greg T., ed. *Water for Southern California: Water Resources Development at the Close of the Century*. San Diego: San Diego Association of Geologists, 1999.

Daley, Beth. "Houses, Resorts Planned for North Woods." *Boston Globe*, 5 April 2005, A1.

———. "Paul Bunyan on Trial." *Boston Globe*, 19 September 2000, D1.

———. "A Race to Save Maine Woods." *Boston Globe*, 18 September 2000, A1.

Dana, Richard Henry, Jr. *Two Years before the Mast: A Personal Narrative of Life at Sea*. Garden City, NY: Doubleday, 1949.

Darnell, William Irvin. "The Imperial Valley: Its Physical and Cultural Geography." Master's thesis, California: El Centro Public Library, 1959.

"Desalination in Florida May Fuel Future Growth." *Growth/No Growth*, September 2002.

Douglas, William A., and Robert A. Nylen, eds. *Letters from the Nevada Frontier: Correspondence of Tasker L. Oddie, 1898–1902*. Reno: University of Nevada Press, 1992.

Douthat, Ross, Nathan Littlefield, and Marshall Poe. "Primary Sources: Spoil the CEO . . ." *Atlantic Monthly*, May 2005, 52.

Dowd, M. J. *Imperial District: The First 40 Years: History of Imperial Irrigation District and the Development of Imperial Valley*. Imperial, CA: Imperial Irrigation District, 1956.

Drache, Hiram M. *The Day of the Bonanza: A History of Bonanza Farming in the Red River Valley of the North*. Fargo: North Dakota Institute for Regional Studies, 1964.

Egan, Timothy. "Drilling in West Pits Republican Policy against Republican Base." *New York Times*, 22 June 2005, All (one-one).

———. "Paying on the Highway to Get Out of First Gear." *New York Times*, 28 April 2005, A1.

El Centro Public Library. "The History and Events of El Centro." 2003. Special Collections, El Centro Public Library, El Centro, CA.

Elliott, Russell. *History of Nevada*. Lincoln: University of Nebraska Press, 1973.

"Executive Pay: What the Boss Makes." *Forbes*, 10 May 2004, 134.

Farr, F. C. *The History of Imperial County, California*. Berkeley, CA: Elms and Franks, 1918.

Greenhouse, Steven. "Report Shows Americans Have More 'Labor Days.'" *New York Times*, 1 September 2001.

Hahamovitch, Cindy. *The Fruits of Their Labor: Atlantic Coast Farmworkers and the Making of Migrant Poverty.* Chapel Hill: University of North Carolina Press, 1997.

Hamer, David. *New Towns in the New World.* New York: Columbia University Press, 1990.

Hine, Robert V., and John Mack Faragher. *The American West: A New Interpretive History.* New Haven: Yale University Press, 2000.

Hinton, Richard J. *Letter from the Secretary of Agriculture Transmitting, In response to Senate resolution of December 13, 1890, a report on the progress of irrigation investigation under the deficiency appropriation act of 1890.* 51st Congress, 2d Session, Senate Ex. Doc. 53, 1890.

"History of Public Land Livestock Grazing." Bureau of Land Management. http://www.nv.blm.gov/range/History_of_Grazing.htm.

Howe, Edgar F., and Wilbur Jay Hall. *The Story of the First Decade.* Imperial, CA: Imperial County Historical Society, 1998.

Hu, Winnie. "To Protect Water Supply, City Acts as a Land Baron." *New York Times,* 9 August 2004, A14.

"Idaho: Loss of Farmland Hurts Local Economies." *Growth/No Growth,* September 2002.

Imperial County Historical Society. "A History of the Imperial Valley." Imperial Valley Pioneers Museum, Imperial, CA. www.imperial.cc.ca.us/Pioneers/.

"The Importance of Being Private." *Forbes,* 29 November 2004, 201.

Jehl, Douglas. "Gold Miners Eager for Bush to Roll Back Clinton Rules." *New York Times,* 16 August 2001.

Johnson, Carrie. "Firm Indicted in Calif. Energy Case." *Boston Globe,* 9 April 2004, A2.

Johnson, Kirk. "The West's New Boomtowns Are Looking Beyond the Drought." *New York Times,* 3 February 2005, All.

Kilborn, Peter. "Boom in Economy Skips Towns on the Plains." *New York Times,* 2 July 2000, A1.

Lee, Jennifer. "7 States to Sue E.P.A. over Standards on Air Pollutioin." *New York Times,* 21 February 2003.

Lincoln, Abraham. "March 9, 1864.—Message to the Senate." In *Complete Works, Vol. II, Letters and State Papers of Abraham Lincoln,* ed. John G. Nicolay and John Hay. New York: Century Co., 1894.

MacKaye, Benton. "Employment and Natural Resources." Washington, DC: Government Printing Office, 1919.

Marshall, Colonel William R. "Journal of Military Expedition Against Sioux, 1863, Under Command of Brig. General Henry Hastings Sibley, 7th Minnesota." Special Collections, Chester Fritz Library, University of North Dakota.

McCarthy, Terry. "High Noon in the West." *Time,* 16 July 2001, 18–32.

Meacham, Bradley. "Preservation Deal for Forest Falls Through." *Seattle Times,* 11 March 2003.

Morgenson, Gretchen. "Explaining (or Not) Why the Boss Is Paid So Much." *New York Times,* 25 January 2004.

Mouawad, Jad. "As Geopolitics Takes Hold, Cheap Oil Recedes into the Past." *New York Times*, 3 January 2005, C4.

Murphy, Dean E. "Pact in West Will Send Farms' Water to Cities." *New York Times*, 17 October 2003, A1.

———. "Water Contract Renewals Stir Debate between Environmentalists and Farmers in California." *New York Times*, 15 December 2004.

Oberly, James W. *Sixty Million Acres: American Veterans and the Public Lands before the Civil War*. Kent, OH: Kent State University Press, 1990.

Parsons, James J. "A Geographer Looks at the San Joaquin Valley." Carl Sauer Memorial Lecture, University of California, Berkeley, 1987. http://geography.berkeley.edu/ProjectsResources/Publications/Parsons_SauerLect.html.

"Private Mission." *Forbes*, 29 November 2004, 204.

"Reclamation Act/Newlands Act of 1902." Center for Columbia River History, Washington State University, Vancouver. http://www.ccrh.org/.

Reform of the Land Laws: Conservation of National Resources/Extracts from Recommendations of the President, The Secretary of the Interior, and the Commissioner of the General Land Office, etc. Senate Document No. 283. Washington, DC: Government Printing Office, 1910.

Reidy, Chris. "As Profits Slip, Inns on Nantucket Are Converted to Homes." *Boston Globe*, 24 May 2004, A1.

Report of the General Land Office. Washington, DC: Government Printing Office, 1901.

Report of the General Land Office 1838. Washington, DC: Government Printing Office, 1838.

Report of the General Land Office 1851. Washington, DC: Government Printing Office, 1851.

Ridout, Christine F. "The New England Desert." *Sanctuary* 12, no. 3 (spring 2000): 9.

Rippy, Brad. "U.S. Drought Monitor." Washington, DC: United States Department of Agriculture, 25 January 2005. www.drought.unl.edu.dm/monitor .html.

Rolfsrud, Erling Nicolai. *The Story of North Dakota*. Alexandria, MN: Lantern Books, 1963.

San Diego Historical Society. "History of San Diego, 1542–1908." www.sandiegohistory.org.

San Diego Water Authority. *San Diego's Quest for Water: The Metropolitan Water District of Southern California*. San Diego: San Diego Water Authority, 1947. Special Collections, San Diego Public Library.

Stephenson, George. *The Political History of the Public Lands from 1840–1862*. Boston: Richard G. Badger, 1917.

Taylor, Paul Schuster. *Essays on Land, Water, and the Law in California*. New York: Arno Press, 1979.

Turner, Frederick Jackson. *The Frontier in American History*. New York: Henry Holt, 1920.

Western Management Consultants. *Water and San Diego County Growth: A Study*

for the San Diego County Water Authority. Phoenix: Western Management Consultants, 1966.
Williams, John Hoyt. *A Great and Shining Road: The Epic Story of the Transcontinental Railroad*. New York: Random House, 1988.
Wilson, Janet. "EPA Fights Waste Site Near River." *Los Angeles Times*, 5 March 2005.

SECTION 5
AMERICANS AND THEIR LAND

American Farmland Trust. "Farming on the Edge." http://www.farmland.org/farmingontheedge/major_findings.htm.
Barboza, David. "America's Cheese State Fights to Stay That Way." *New York Times*, 28 June 2001, C1.
Becker, Elizabeth. "Far from Dead, Subsidies Fuel Big Farms." *New York Times*, 14 May 2001, A1.
———. "Prairie Farmers Reap Conservation's Rewards." *New York Times*, 27 August 2001.
Berenson, Alex. "For Merck, the Vioxx Paper Trail Won't Go Away." *New York Times*, 21 August 2005, A1.
Blayney, Don C., and Alden C. Manchester. "Large Companies Active in Changing Dairy Industry." *FoodReview* 23, no. 2 (2000): 1–2.
"A National Milk Plan." *Boston Globe*, Editorial, 11 November 2001.
Carter, Bill, and Jim Rutenberg. "Creative Voices Say Television Will Suffer in New Climate." *New York Times*, 3 June 2003, C1.
"Changes in the Way Political Campaigns Are Financed." *New York Times*, 4 March 2002, A16.
Daley, Beth. "Handlers Press for a Hike in Bottle Fees." *Boston Globe*, 31 January 2002, B1.
Dao, James. "Rule Change May Alter Strip-Mine Fight." *New York Times*, 26 January 2004, A12.
Department of Justice. "Justice Department Files Lawsuit to Block Suiza Foods Corporation's Acquisition of Broughton Foods Company." Press Release, 18 March, 1999. www.usdoj.gov/atr.
"Federal Judge Eases Snowmobile Curbs." *New York Times*, 12 February 2004.
Gardner, Howard. *Frames of Mind: The Theory of Multiple Intelligences*. New York: Basic Books, 1985.
"The Importance of Being Private." *Forbes*, 29 November 2004, 201.
Kaufman, Marc. "Poll: Americans Value Drugs, Question Makers." *Washington Post*, 25 February 2005.
Kirkpatrick, David D. "New Rules Give Big Media Chance to Get Even Bigger." *New York Times*, 3 June 2005, C1.
Labaton, Stephen. "Regulators Ease Rules Governing Media Ownership." *New York Times*, 3 June 2003, A1.
Leahy, Patrick. "Leahy Unveils Comprehensive Agriculture Antitrust Bill." Press Release, 12 April 2000.

McKee, Bradford. "Growing Up Denatured." *New York Times*, 28 April 2005.

Murray, Barbara. "Dean Foods Company." http://www.hoovers.com/dean-foods/—ID_42651—free-co-factsheet/xhtml.

Ocasso, Carla. "Farmer Sends Manure Instead of Milk." *Caledonian-Record News*, October 2002, 23.

O'Leary, Wayne. "What Scandal? It's Business as Usual." *Maine Telegram*, 21 July 2005, C1.

"Private Mission." *Forbes*, 29 November 2004, 204.

Rimer, Sue. "Sisters Give Up the Cows and Gain Richer Lives." *New York Times*, 28 August 2001.

Robinson, Sue. "Vt. Farmers Fear Suiza Domination." *Burlington Free Press*, 25 June 2000, 1A.

Rogers, Will. "It's Easy Being Green." *New York Times*, 20 November 2004.

Rothstein, Richard. "Schools' Chosen Cure for Money Ills: A Sugar Pill." *New York Times*, 21 August 2002.

Saul, Stephanie. "Making a Fortune by Wagering That Drug Prices Tend to Rise." *New York Times*, 26 January 2005, A1.

Schlosser, Eric. *Fast Food Nation.* Boston: Houghton Mifflin, 2001.

Seelye, Katharine Q. "E.P.A. and Energy Department War over Clean Air Rules." *New York Times*, 19 February 2002.

———. "Snowmobilers Gain against Plan for Park Ban." *New York Times*, 20 February 2002, A14.

"Snowmobiles Unbound." *New York Times*, Editorial, 14 February 2004, A28.

Staples, Brent. "The Trouble with Corporate Radio: The Day the Protest Music Died." *New York Times*, 20 February 2003, A30.

Stollberg, Sheryl Gay, and Gardiner Harris. "Industry Fights to Put Imprint on Drug Bill." *New York Times*, 5 September 2003, A1.

"Suiza Foods Settles New England Antitrust Probe." *Dallas Business Journal*, 25 June 2001. http://www.bizjournals.com/dallas/stories/2001/06/25/daily11.html.

"Suiza to Divest More Plants in Dean Foods Merger." *Dallas Business Journal*, 6 December 2001. http://www.dallas.bizjournals.com/dallas/stories/2001/12/03/daily44.html.

Van Natta, Don, Jr., and Neela Banerjee. "Top G.O.P. Donors in Energy Industry Met Cheney Panel." *New York Times*, 1 March 2002, A1.

Wayne, Leslie. "Enron, Preaching Deregulation, Worked the Statehouse Circuit." *New York Times*, 9 February 2002, B1.

Wells, Nancy M. "At Home with Nature: Effects of 'Greenness' on Children's Cognitive Functioning." *Environment and Behavior* 32, no. 6 (November 2000): 775–95.

———. "Nearby Nature: A Buffer of Life Stress among Rural Children." *Environment and Behavior* 35, no. 3 (May 2003): 311–30.

INDEX

Text design by Jillian Downey
Typesetting by Delmastype, Ann Arbor, Michigan
Text font: Janson
Display font: Bank Gothic BT

Although designed by the Hungarian Nicholas Kis in about 1690, the model for Janson Text was mistakenly attributed to the Dutch printer Anton Janson. Kis' original matrices were found in Germany and acquired by the Stempel foundry in 1919. This version of Janson comes from the Stempel foundry and was designed from the original type; it was issued by Linotype in digital form in 1985.

—courtesy www.adobe.com

Bank Gothic BT was designed at American Type Founders in 1930–33 by Morris F. Benton.

—courtesy www.myfonts.com